W9-BSC-453

British Ferns

British Ferns

RON FREETHY

The Crowood Press

First published in 1987 by
The Crowood Press
Ramsbury, Marlborough,
Wiltshire SN8 2HE

British Library Cataloguing in Publication Data

Freethy, Ron
British ferns.
1. Ferns – Great Britain
I. Title
587'.31'0941 QK527

ISBN 0 946284 33 4

Picture Credits

Line illustrations by Carole Pugh

Typeset by Inforum Ltd, Portsmouth
Printed in Great Britain
at the University Printing House, Oxford

Contents

Introduction

There are many technical books on the biology of ferns. This is not intended to be another. Its aim is to provide a simple introduction to British ferns and their close allies for the amateur. It had its origins in a WEA university extension course for naturalists. One retired lady asked if there was a book just about the ferns of Britain, giving details of their folklore and habitat in addition to simple descriptions. To my surprise there was not, and this book has been written to fill the gap. It is dedicated to 'Nellie Fern', the retired lady who wanted to know and who inspired this book.

I am also grateful to a number of others, especially Marlene Freethy whose skill on the typewriter is appreciated as much as her botanical knowledge. Brian H. Lee read the manuscript and kept in touch during the gestation period, and his nose for detail was, as ever, appreciated. Although correctly acknowledged elsewhere in the book, I must mention my gratitude to Carole Pugh for working so hard on the illustrations. Finally, a thank you to a number of students who read sections and asked questions which enabled me to clear up points of confusion. The errors which do remain are, of course, my sole responsibility.

1 Ferns in the Plant Kingdom

If we walked into a British woodland in late spring, the first thing we should notice would be beautiful wedges of colour as the yellow primroses and white wood anemones gave way to the waving blue ranks of bluebells. In the deep recesses, the very light would have a blue mist about it and the sweet smell of blossoms would caress the senses. Only if we looked closely on the shady banks would we begin to notice the ferns. The bracken would be beginning to unfold, and in its youth would look surprisingly like a shepherd's crosier. In woodlands growing on limestone, hart's tongue fern would be peeping out between the cracks, but in every case the ferns would be playing a subservient role, overshadowed by trees such as oak, ash, beech, elm and several other giant species.

How different things would be if we could transport ourselves backwards in time to the Carboniferous age which began some 350 million years ago, when ferns were the dominant form of vegetation, often growing to heights of up to 15.25m (50ft) whilst their close relatives the club-mosses and the horsetails grew almost twice as high. Some botanists think that the comparatively smaller ferns evolved their large, finely divided leaves so that they could soak up the maximum amount of light.

The conditions under which they grew best tended to be warm and wet, and as the plants blew down in storms they became crushed beneath the weight of those which fell later. Over the centuries these plants became further compressed into coal, only to be disturbed as insatiable human demands for cheap sources of energy developed during the Industrial Revolution. If you are reading this book in front of a coal fire, look into the flames and feel the warmth of the energy which arrived in waves from the sun all those years ago, and has been locked within the body of the giant fern ever since.

The age of the fern, however, has long since passed and the species which ended up in the coal measures are now extinct. Flowering plants are now the dominant form of vegetation. Nevertheless there are still more than ten thousand ferns, horsetails and club-mosses on earth today, although only a tiny fraction are found in Britain. Evolution is a continuous process, and the flowering plants will no doubt develop innovations which will make them more efficient. The ferns have long been overtaken on the evolutionary road, but it is fascinating to see what a giant step forward they made all those millions of years ago.

Evolution of Plants

The plant kingdom is conveniently divided into two groups: the flowering plants, called the phanerograms, and the non-flowering plants, called the cryptograms. The cryptograms were so named because when early botanists hunted for signs of flowers they found none and thought they were hidden away in a 'crypt'. The non-flowering plants are divided into three important phyla, namely algae, liverworts and mosses, and finally ferns and their allies which are the most advanced plants belonging to the cryptograms. A true understanding of how ferns evolved and why they have now been left behind can only be fully attained through a brief account of each of the phyla making up the cryptograms.

The Thallophyta

This is the phylum which includes the algae, and until recently taxonomists also included both bacteria and fungi within the thallophyta. I have followed what I think is the accepted policy by only including the algae in this phylum, the bacteria and fungi now being deemed worthy of classification on their own.

It seems certain that the earliest plants (and the animals which fed upon them) lived in water and were totally incapable of survival out of this element. All living tissues consist mainly of water, and if this vital liquid is absent then the organism will die. On land the tendency is for the body to dehydrate, especially in warm open spaces. During evolution organisms seem to have moved from shallow water to damp land, and then to dry land. This has taken millions of years and has required many significant structural alterations. The algae, of which the best modern examples are the seaweeds, are almost totally aquatic and have been able to keep their structural development simple. The water gives such buoyancy that no supportive tissue is required, even in the largest species.

We should not leave the thallophyta without a brief mention of a most remarkable example of mutual co-operation between algae and fungi which goes under the name of lichen. Algae, being green plants, are able to produce food from sunlight, but find survival on land difficult. Fungi, on the other hand, are basically land-based organisms and they are not able to photosynthesise. They are, however, adept at anchoring themselves to rocks. The two organisms live together to their mutual benefit, the algae being the food provider and the fungi the sheet anchor. Biologists call this co-operation symbiosis, and it does seem to be a cul-de-sac off the main evolutionary road.

The Bryophytes

This phylum includes the liverworts and mosses which are found in very damp conditions, although they have at least broken free from total dependence on aquatic habitats. They have in fact evolved several strategies to make life possible on land. Arguably the most important advance is the development of primitive leaves, although they consist of only one layer of cells. These are covered in the Polytrichum mosses, with a layer of waxy material called lamellae which cuts down the rate at which water evaporates from the surface. This ability to retain some water enables bryophytes to withstand very short periods of drought.

The second and perhaps equally important advance is in the development of rhizoids. These look rather like the roots of the higher plants, which have three vital functions. The plant is anchored to the ground by its roots, water is absorbed from the soil and transported up to the leaves, and finally sugar is transported back from the leaves, where it is made during photosynthesis, to the roots. The rhizoids of bryophytes only carry out the first two of these functions and are therefore not classified as true roots. They work rather like a strip of blotting paper, allowing the liquid to creep up by the process of capillarity. What the bryophytes have totally failed to do is to develop any form of supportive tissue, and they must rely upon having their cells full of water in order to stand upright. There are no vessels to transport water (called xylem) or others to transport sugar (called phloem) and this has restricted their exploration of land to the damp edges of streams and under wet hedges and banks.

Bryophytes are divided into the Hepaticae or liverworts and the Musci or mosses.

The liverworts are an ancient group and fossils have been found in the coal measures. They grow mainly in damp environments: by the edge of streams, in and around swamps, and a few species still grow in water. The plants are almost always flat and lie close to the ground, a condition referred to as dorsiventral. A few more advanced species in the order Jungermanniales (foliose liverworts) have developed a thin weak stem which bears primitive leaves, but the Hepaticae are still a long way from overcoming the problems of living successfully on land.

In the mosses, whose fossils take us back to the Carboniferous era, there are signs of more efficient mechanisms for coping without water. The body has a stronger stem and more highly evolved leaves than the liverworts. If the stem is examined under a microscope there are signs of strengthening tissue but no true xylem or phloem, and cellulose, which is the typical material making up the cell walls of higher plants, only occurs in a small minority of mosses. It is not the purpose of this book to describe the bryophytes, but some understanding of their reproduction is essential if we are to place what has happened to the ferns during their evolution in perspective.

Life Cycle of a Moss

The basic process is the same for all mosses, but there will be minor variations from species to species. *Mnium hornum* is a common woodland species and also grows well along the banks of streams. The leafy stems are usually about 2.5cm (1in) long and 0.8cm (0.3in) wide, and if the ends of some stems are examined in May a number of brown hair-like structures can be seen protected by a rosette of leaves. These are the male plants and at the base of the brown 'hairs' the antheridia are seen. These produce the male cells which are able to swim through a thin layer of water, being attracted towards the female plants by a chemical which they produce.

Female plants may be recognised by a terminal cluster of pointed leaves, in contrast to those of the male plant which are cup shaped. Some of the leaves on the female plant appear to be darker green and longer than others and these enfold and protect the female organs which are called

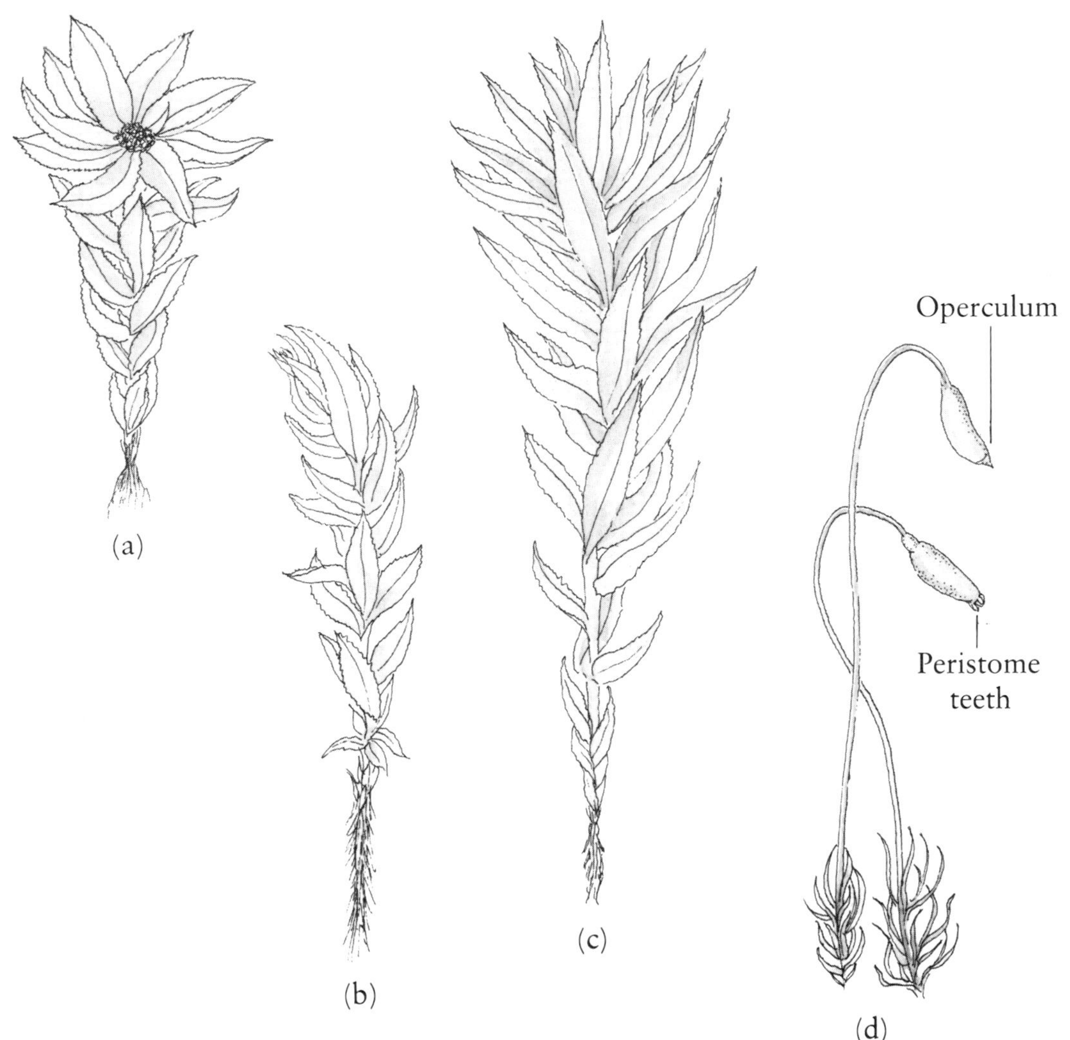

The moss, *Mnium hornum*. (a) Male. (b) Vegetative. (c) Female. (d) Shoots with capsule (after Priestley and Scott).

archegonia. Each archegonium has a huge stalk-like structure called the venter, narrowing to a neck filled with cells which, when the structure is ripe, break down into mucilage. The male cell (called a gamete) is attracted towards the neck of the archegonium and swims through it to fertilise the female cell. Since the moss plant at this stage produces sexual gametes it is called the gametophyte generation.

The fertilised oospore then proceeds to develop a conical structure called the foot, which buries itself in the body of the gametophyte plant and lives upon it as a parasite. Using the energy derived from the parent, a stalk-like structure up to 5cm (2in) long, called the seta, develops quickly. A swelling occurs at the end of the seta and is called the capsule. Inside the capsule the spores develop. As the capsule dries in the summer sun, a lid called the operculum bursts open to release the spores. Because this moss plant produces only non-sexual spores it is called the Sporophyte generation. This type of life history, well known to biology students sitting examinations, is known as the alternation of generations. It is clear that the gametophyte generation cannot fulfil its function in the absence of water, but the sporophyte generation is perfectly able to manage without water – indeed, the spores will be dispersed more efficiently in dry weather. The spores, however, do not need a moist environment in which to germinate. In the early stages the new gametophyte plants look like a green filament, but the typical moss plants are soon recognisable.

The Pteridophytes

In the life cycle of the fern there is also a clearly recognisable alternation of generations, the gametophyte stage being followed by the sporophyte stage. In the ferns, however, there has been a reversal in the relative dominance; the large leafy sporophyte is no longer parasitic upon the gametophyte, but the major life form. The gametophyte is reduced to a tiny, but nevertheless vital, plate-like structure with the sole function of producing mobile male gametes which still need to swim through water to reach the female gamete. Because the sporophyte is much less dependent on water, the ferns have been able to advance further along the road to life on dry land.

It is not only ferns which make up the thallophytes, and it is interesting to see how the class as a whole has placed a gradually increasing emphasis on the importance of the sporophyte generation at the expense of that of the gametophyte. Thallophytes are divided into four subclasses: quillworts, club-mosses, horsetails and ferns. We must be careful not to over-simplify, since the taxonomy is complicated, and the situation is not helped by the large numbers of fossil forms involved. This book, however, is only concerned with modern ferns and also with a few of their living close relatives.

The pteridophytes vary a great deal in form, but two main evolutionary lines can be detected. Some have leaves which, compared to the stem, are tiny and are therefore called microphyllous pteridophytes. The club-mosses and horsetails clearly belong to this group. The ferns, on the other hand, have very large leaves (or fronds) compared to the stems and are therefore referred to as megaphyllous pteridophytes.

In Chapters 2 and 3 I will discuss the microphyllous pteridophytes, and in Chapter 4 the structure and life history of the ferns will be described in detail. The rest of the book will then consider the habitats in which British ferns occur, although it is realised that some ferns such as the bracken are so widespread that they could be described under almost any habitat heading.

2 Quillworts and Club-mosses

Quillworts

Quillworts are quite small perennial plants, which although occasionally terrestrial are mostly aquatic, and all the modern species are placed in one order called the Isoetes. Isoetes derives from two Greek words meaning evergreen, and indeed these plants do remain in leaf throughout the winter. Although very insignificant to us, the quillworts have a proud and ancient lineage going back to the Carboniferous period when the tree-like Sigillaria was dominant in some areas and has been found as fossils in the coal measures.

Only two species of quillwort are usually found in Britain, namely *Isoetes lacustris* and *Isoetes echinospora*. At one time they were called pepperworts. The former is mainly confined to mountainous regions and this means that in Britain the plant has a definite bias in distribution towards the north and west. Wales, the English Lake District and the mountains of Ireland and Scotland are good spots to hunt for the quillwort, where shallow lakes and ponds with bottoms of clay, sand or gravel are suitable habitats. The depth at which Isoetes grows can vary from a few centimetres to several metres, but it depends upon water clarity rather than any other factor because, as in all plants, light is essential for the process of photosynthesis. In ideal conditions quillwort can be so prolific that it resembles an underwater meadow.

Isoetes echinospora is a smaller, more delicate species which grows in lakes with peaty bottoms. The two species produce spores which differ sharply in appearance, although a simple microscope is necessary to see this. The spores of *Isoetes lacustris* are yellow and covered with roundish spots, whilst those of *echinospora* are white and, as the specific name indicates, covered by tiny brittle spines. This species has a more southerly distribution, extending into Cornwall, and is particularly common in the peaty pools on Bodmin Moor in the vicinity of Jamaica Inn.

There is also a third species, *Isoetes hystrix* which is a terrestrial and Mediterranean based plant, but which has been recorded both in the Channel Islands and at the Lizard point in Cornwall. All the quillworts represent something of a problem to the inexperienced observer in that their leaves show a great similarity to sedges and rushes, which tend to grow in the same habitats. Altogether there are around sixty species, almost entirely distributed in the Northern hemisphere and *Isoetes lacustris* is typical of the genus in both structure and mode of reproduction.

Structure

The plant grows from an underground (or perhaps we should say undermud) stem called a rhizome which is fleshy and about the same size as a hazel nut. It is frequently flattened and used as a food store which means that it should be classed as a corm. As the rhizome ages it branches into two lobes which point downwards and are between 2.5 and 3.75cm (1 and 1.5in) long. From this structure a straggle of root-like structures called rhizoids penetrate the bottom of the pond and serve as a firm anchor, often reaching depths of 10cm (4in). The tufts of translucent green leaves sprouting from the corm do indeed look quill-like, and when mature can reach heights of more than 15cm (6 in). A good specimen may have as many as seventy leaves which, although very firm at the base, hang loosely from the tip, giving the quillwort the appearance of an opened fan.

Each leaf has a ligule which is a strap-like strip of tissue at the base of the leaf. This is also a feature of many grasses and is one more reason why the presence of quillworts is often not noticed by naturalists. If a leaf is examined against the light, many interior transverse partitions can be detected which do not occur in flowering plants, whilst a close look at their bases will confirm the presence of a sporangium buried in the leaf tissue. Examinations over a whole growing season will reveal the presence of two types of sporangia, and will also confirm the close relationship between quillworts and the ferns, and the quillwort's descent from a primitive bryophyte ancestor.

The leaves growing during the early part of the year carry megasporangia which produce many megaspores. These are situated in the angle between the ligule and leaf. Each megaspore grows into a female plant called a prothallus which has on it archegonia containing fertile female gametes. The fertile female gametes take time to mature and so, later in the season, have at their bases microsporangia which contain even larger numbers of microspores, each of which can develop into a male prothallus. These bear antheridia which produce mobile spermatozoids, the male gametes that are chemically attracted to the archegonia. From the fertilised 'egg' a new sporophyte generation develops whilst the 'evergreen' parent carries on growing until well into autumn, the leaves produced during this period being sterile. The food produced by the leaves is passed back into the corm-like rhizome and thus enables quillworts to be perennial even though the alternation of generations type of life cycle is still clearly evident.

Club-mosses

The club-mosses do indeed bear a superficial resemblance to mosses, but a close examination of the stem will reveal the presence of vascular (strengthening) tissue which has not been evolved by mosses. The process of reproduction is also less like that of the moss and closer to that employed by ferns. This has led some botanists to propose the club-mosses as an intermediate stage between mosses and ferns, but it would be far safer to say that this route is just one of many possible evolutionary pathways and could well be another blind alley. What we can say with certainty is that the club-mosses are a fascinating group and one which should be examined by all students of ferns. Once much more widespread and important,

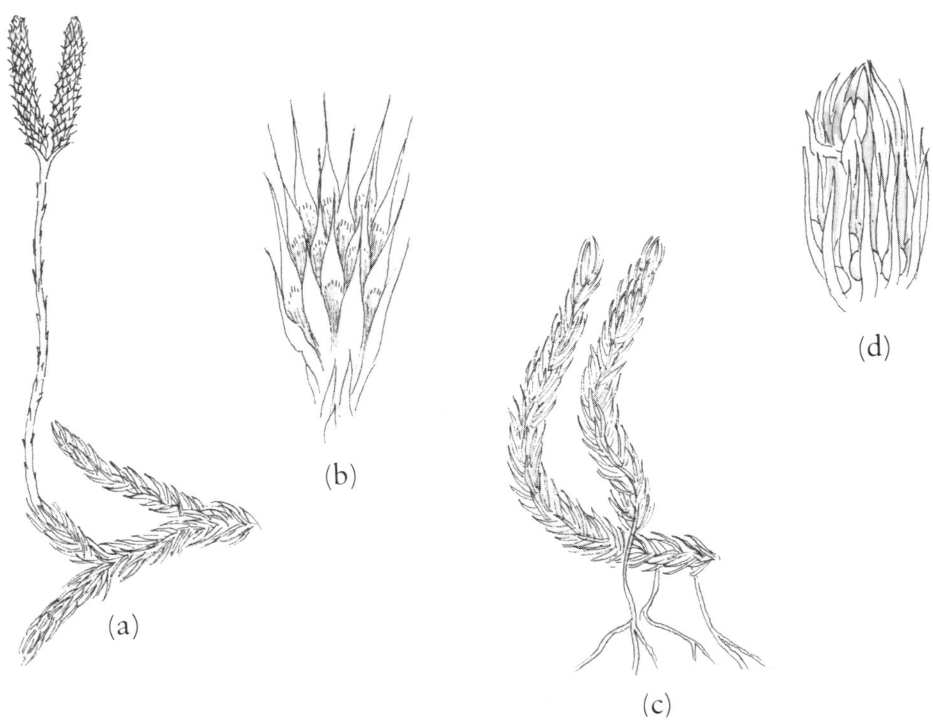

The club-moss, Lycopodium. (a) Normal growth pattern of most species. (b) Sporophyll with sporangium. (c) *Lycopodium clavatum*. (d) Sporophyll with sporangium.

there are now only two genera – the Phylloglossum, which are only found in Australasia, and the Lycopodium. The latter consists of two families: the Lycopodiaceae, which has five British species, and the Selaginellaceae which is basically tropical and includes around seven hundred species. Only the prickly club-moss *Selaginella selaginoides* occurs wild in Britain.

The genus Lycopodium, formerly known as Selago, is made up of around one hundred species, the British representatives being particularly common on wet heathlands, primarily on those at high altitudes in North Wales, the Lake District and the Highlands of Scotland. The five British species are the common club-moss or wolf's claw *Lycopodium clavatum*; the fir club-moss *Lycopodium selago*; the alpine or savin leaved club-moss *Lycopodium alpinum*; the interrupted club-moss *Lycopodium annotinum*; and finally the marsh club-moss *Lycopodium inundatum*. The wolf's claw and the fir club-moss are the commonest species and the account which follows is based largely upon a study of their physical appearance and life history. Recent attempts to reclassify the club-mosses will be discussed at the end of the chapter.

Biology

The naked rhizome (underground stem) puts out branches which creep along the ground and are thickly clustered with spirally arranged leaves, a typical microphyllous arrangement. Branching at the axis of the stem does occur, but it is not a simple branching into two. It is described as monopodical, which means that the apical bud on the main axis always remains dominant. Thus the plant can have branches rather like a tree, a much more advanced arrangement than that in primitive plants, where the apical bud branches dichotomously. In these plants there are two main branches, and the initial line of growth has been lost.

There are adventitious roots on both the rhizome and the creeping surface stems which serve to anchor the plant. The lance-shaped leaves end in a sharp point which makes them quite prickly to the touch. By looking at the leaf margins it is possible to distinguish between *Lycopodium clavatum*, which have serrated edges, and those of *Lycopodium selago* which do not. There is also an important difference to be seen when the two club-mosses are fertile. The wolf's claw produces erect, aerial branches which are called podia and are lightly covered by structures resembling small scales. At the top of the branches, which are usually forked, is a pair of cone-like structures called strobili. These are constructed of specialised leaves termed sporophylls and on the inside surface of these are the reproductive parts. The sporophylls are much smaller and more closely crowded in a spiral around the stem than the vegetative leaves.

In *Lycopodium selago*, the fir club-moss, the fertile parts are more difficult to distinguish from the rest of the plant, and the sporophylls are almost identical to the normal leaves. Specialised leaves are produced to protect the strobili, but they are followed in succession by ordinary leaves and thus there are alternating regions along the stem of sterile and fertile leaves. These are never found in *Lycopodium clavatum*. In addition to its normal method of reproduction, the fir club-moss has a vegetative method of spreading. Just behind the apices of the young shoots there often occur detachable swellings, called gemmae or bulbils, which quickly fall to the ground and produce new plants.

Alternation of Generations in *Lycopodium clavatum*

As we have seen, each strobilus is made up of tightly spiralled sporophylls arranged around a central axis. The kidney-shaped sporangia is carried singly on the inner surface (called the adaxial side) of the sporophylls. These sporophylls are unlike the vegetative leaves in having a broad base which is overlapped to protect the sporangia. They also contain very little chlorophyll, and their nutritional value must be regarded as minimal compared to the vegetative leaves. The whole structure bears a great resemblance to the cones of a larch or Scots pine.

The sporangia are carried on short stalks and each is covered by a three-cell-thick wall, the inner one, called the tapetum, acting as a food store for the archesporial tissue within. This gives rise to four spores, and constitutes the asexual part of the reproductive process. The strobilus is therefore the sporophyte generation and each gives rise to many sporangia with its own quartet of spores. These have been harvested for many years and marketed under the name of lycopodium powder. Physicists used it in experiments on

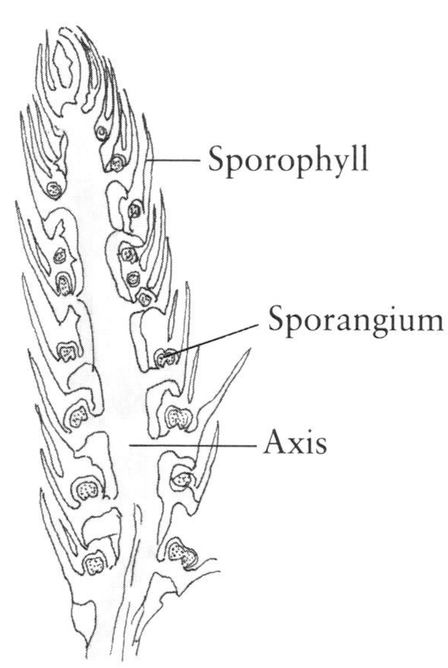

Cone structure of *Lycopodium clavatum*.

sound, the lightness of the spores being useful in detecting the slightest of vibrations. All the spores look alike and all the Lycopodium species are therefore said to be homosporous.

The work of a naturalist is not always easy and it took a great deal of patient work to discover the next stage of the life history, which is the development of the gametophyte generation. The difficulty was that Lycopodium spores tend to remain dormant for several years, and even then the gametophyte can take several more years to mature. Two types of gametophyte or prothallus have been recognised. Some do develop very quickly and produce chlorophyll and then fertile offspring within one year. Of the British species only *Lycopodium inundatum* develops this quickly. Of the second type, the period between spore production of the sporophyte and fertile embryos being produced by the gametophyte can be between twelve and twenty years. To add to the difficulty, the sporophyte which develops from the embryo is often colourless, subterranean and lives as a parasite on the body of the gametophyte. This stage may last a further ten years! Both *Lycopodium clavatum* and *Lycopodium selago* have this type of extremely complex alternation of generations.

Identifying Club-mosses

As may be imagined, with a life history as complex as this the club-mosses are impracticable to cultivate and the real attraction is to find them growing in the remote areas of Britain surrounded by wild and often savage scenery. The five species found in Britain do show habitat preferences and it is possible, with practice and patience, to distinguish them physically. *Lycopodium clavatum*, the wolf's claw or common club-moss is, like the rest of the genus, an alpine species and although it is found at very high altitudes in the tropics, it is mainly confined to the hills of the Northern hemisphere as well as throughout the Arctic. In Britain it favours the highland areas of the North and West, but it was once found frequently in the South. Its decline from heaths and moors in the South is probably more due to loss of habitat than climate. Knowing that the species has an interesting life history, many tourists who should know better remove it and try to keep it; 'a throw back to the stone age', as I heard one educated vandal describe it. Before the modern electronic age, the strobili were crushed to produce a highly inflammable powder which could be used to create artificial lightning on the stage of theatres. The yellow spores were also once in demand by pharmacists for coating pills. It is the long stalked strobili, in addition to the long creeping stems, which distinguish this from the rest of the genus. The specific name of clavatum means club-shaped, and is certainly a good description of the cone-shaped strobili.

Lycopodium selago, the fir club-moss is very rare in lowland districts and has not been recorded in southern England for many years. It is a true montane species and inhabits the summits of most of Britain's highest mountains and those of other Northern hemisphere countries. It is a frequent member of the flora of the arctic tundra. It is unusual in branching dichotomously, a rather primitive arrangement abandoned by the rest of the Lycopodium species. The usual club-moss characteristic of bearing pairs of terminal strobili has also been abandoned.

Lycopodium alpinum, the arctic or savin-leaved club-moss, derives its last name because of the similarity of its leaves to the shape of those of the juniper, *Juniperus communis*, which was also known as the savin. Here, as its name 'alpinus' clearly implies, is a true mountain plant, and in the Highlands of Scotland I have frequently encountered it at heights of almost 1,219m (4,000ft). Its favourite habitat is on grassy but quite exposed slopes, often in the company of the common club-moss. It can be distinguished from the latter, however, by its leaves which are arranged in four rows, and by its distinctive bluish colouration. A further peculiarity often present is a lop-sided appearance due to the leaves on one side of the stem being much larger than those of the other.

Lycopodium annotinum, the interrupted club-moss, is another basically highland species not often occurring below 305m (1,000ft) and often at three times this height on exposed areas of moorland and mountain, especially in Scotland. It appears to be much less common in former haunts in the Lake District, Yorkshire Dales, Derbyshire and North Wales. It is an easily distinguished species because of its lack of pointed leaves and the jointed nature of its stems which make it resemble a string of sausages. Each joint can be almost 2.5cm (1in) long and represents one year's growth. Both the scientific name of 'annotinum' and the English name of 'interrupted club-moss' describe these jointed branches, and another unique feature is that the strobili do not have stalks, a condition termed sessile.

Lycopodium inundatum, the marsh club-moss, occurs both in Europe and in North America. In Britain it is widespread, but rather local and is unusual in the sense that it is more common in lowland areas than on the hills and mountains. The favoured habitat would appear to be marshy land, as the common name of the species implies and it is also particularly attracted to areas subject to regular flooding, especially in winter. This preference is reflected in the specific name of 'inundatum'. Marsh club-moss is often not recorded when a survey is being carried out because it usually grows among a mass of mosses and is not recognised as a club-moss. Once found, the untoothed and curved leaves, and cones borne singly instead of in pairs immediately distinguish it from the other club-mosses.

At one time the fascinating *Selaginella selaginoides*, the prickly club-moss, with its slender straw-coloured stems only reaching 7.5cm (3in), was named *Lycopodium selaginoides*. Many aspects of its anatomy and life history, however, clearly

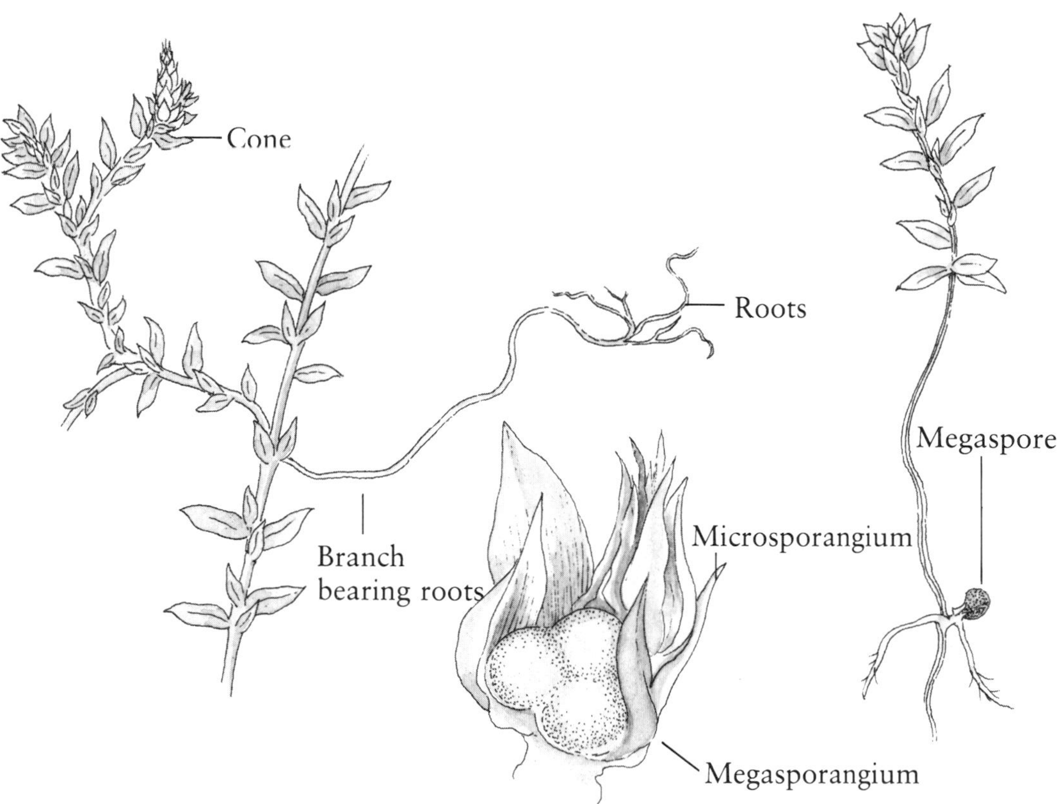

Selaginella kraussiana (after Priestley and Scott).

show it to belong to the tropical family Selaginellaceae. It is a native of damp mountain pastures in the north and west of Britain. A second species *Selaginella kraussiana*, which is a native of Africa, has become a very popular and common greenhouse plant in Britain. Like the rest of the club-mosses the life cycle of *Selaginella selaginoides* has proved very difficult to untangle and therefore most of the research has been carried out on the South African species. The brief account which follows is based upon *Selaginella kraussiana*, but it does seem to be remarkably similar to that found high on the breezy hills of Northern Britain where the prickly club-moss still hangs on to what is probably a diminishing habitat.

Life Cycle of *Selaginella kraussiana*

The sporophyte consists of a creeping stem which appears to be more primitive than it is. The stem branches grow at the same rate as the main axis and the plant therefore looks to be dichotomous whilst it is in fact monopodial. Four rows of leaves are produced in pairs, one being larger than the other, and develop in a horizontal plane giving the impression that the plant has been pressed. The larger leaves arise from the lower lateral surface whilst the smaller ones arise from the upper lateral surface. Each leaf bears an outgrowth called a ligule which is obvious in young plants but gradually withers away. This structure is also obvious in some fossil members of the Lycopodiales which taxonomists suggest show a close relationship to Selaginella.

Just behind each branch in the stem there develops a root-like structure which may or may not reach the soil since it is never more than a few millimetres in length. It puts out a series of smaller 'roots', forming a structure called a rhizophore. There is no similar structure anywhere, and it has been labelled as an organ '*sui genesis*' which means that it has no obvious ancestor. It is a most useful structure to the biologist struggling to identify Selaginella. When mature and fertile, the plant puts up vertical branches which carry the strobili, each of which contains four vertical rows of equal sized sporophylls. Sporangia of two types are developed in the axils of the sporophylls. Some sporangia produce large numbers of small spores called microspores, whilst others give rise to only four large megaspores. Both types are found on the same strobilus but there is a tendency (although no hard and fast rule) for the microsporangia to be situated close to the tip of the strobilus. Thus Selaginella is heterosporous, whereas the Lycopodium club-mosses and ferns are homosporous, producing identical spores.

The two different kinds of spore develop into two distinct types of gametophyte, the microspores becoming male and the megaspores female. There is only one cell division before the microspore is expelled from the sporangium and this isolates a minute cell, which is the only structure to represent the prothallus body. The second cell divides to produce a protective wall of eight cells and four sperms, each consisting of an elongated body containing a nucleus and two whip-like flagellae which drive the structure through any thin film of water towards the female. The microspores are dispersed by the wind, but the megaspores are loaded with food and are therefore a great deal heavier. They tend to fall to the ground close to the parent. Development starts whilst still within the sporangium. The original single nucleus undergoes a number of divisions to produce a partitioned structure containing the reserves of starch laid down in the megaspore, and four female cells. Eventually the megaspore ruptures and the prothallus protrudes through the cracks. It is, of course, quite tiny, although a very substantial structure compared to the male. The archegonia are embedded in the tissue, but a short neck leads to the surface and the mobile sperms are drawn in by chemical attraction. The fertilised oospore enters into a short period of rest before growing for a while on the tissues of the female gametophyte. It then develops chlorophyll and the dominant sporophyte is produced. The method of reproduction followed here, although an alternation of generations, shows great differences from both the bryophytes and other pteridophytes. In the mosses, the sporophyte is dependent upon the gametophyte, and in both the Lycopods and the ferns the gametophytes are nutritionally self-sustaining.

The final word in this chapter must oncern itself with the recent tendency to separate the marsh club-moss from the Lycopodium genus and rename it *Lycopo-*

diella inundata. Likewise the fir club-moss, once *Lycopodium selago*, is also sometimes placed in a genus of its own and named *Hupersia selago*.

3 Horsetails

The club-mosses, horsetails and ferns are described as coevals, by which we mean that they thrived together during the Carboniferous age, and although they obviously had many differences, there were often superficial resemblances. It is because of these similarities that the modern student of ferns would be ill-equipped without a knowledge of their close relatives. Club-mosses often had tree-like representatives up to 30.5m (100ft) high, horsetails grew almost as high and many ferns reached over 15.25m (50ft). There was also another group called the seed-ferns – the Pteridosperms – which are now extinct. The horsetails, Microphyllous Pteridophytes, one of the most primitive groups of plants alive at the present time, belong to the Sphenopsida group which can be divided into two. The order Sphenophyllales includes one family – the sphenophyllaceae – which were plentiful during the Carboniferous era but are now only known as fossils. The second order is the Equisetales containing two families, the extinct Calamitaceae and the Equisetales, which has only one genus called Equisetum. These are the horsetails. The name derives from the latin *equus* meaning a horse and *seta* meaning a bristle.

Identifying Horsetails

All horsetails are typified by hollow jointed stems and tiny, simple leaves arranged in whorls. The sporangia are carried in groups attached to a common stalk called a sporangiophore. These structures are also arranged in whorls, forming terminal cone-like organs called strobili. Twelve species are recognised in Britain but several are not at all common. Confusion often arises because many species have a strong tendency to hybridise and produce plants with intermediate characteristics. The common, or field horsetail is, however, both widespread and prolific.

Field Horsetail

Equisetum arvense, the field horsetail, occurs in waste places especially where there are areas of gravel. The dark brown underground rhizomes are long slender structures covered with delicate reddish hair-like threads. The fact that the rhizomes spread so far and penetrate so deeply into the soil means that they are very difficult to eradicate and are the bane of many gardeners. The rhizomes have been found at depths of almost 3.7m (12ft). At the joints in the rhizome, groups of adventitious roots arise and can both absorb water and increase the efficiency of the anchorage. Some branches remain much shorter than normal and become swollen with food reserves, rather like small potatoes. If these tubers become detached they grow into new plants, yet another characteristic of horsetails well known to horticulturalists who wage war

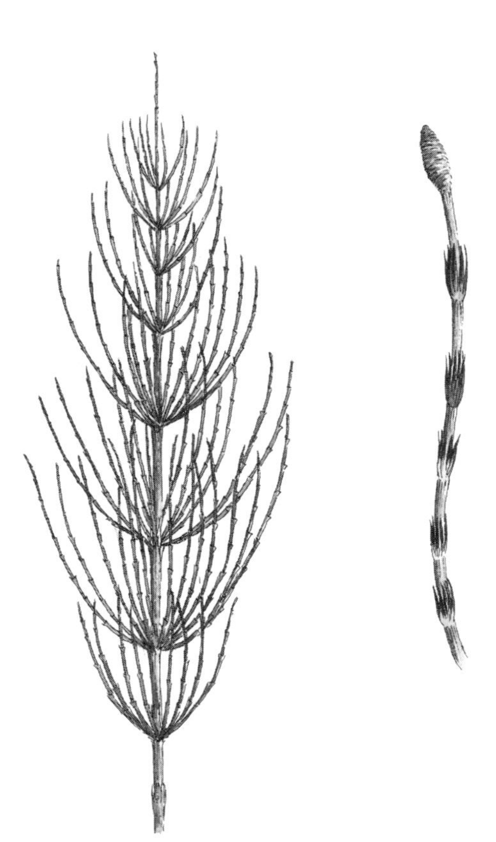

Field horsetail.

against their spread.

The stems arising from the rhizomes are upright, jointed structures bearing whorls of very poorly developed leaves, as could be expected since horsetails are microphyllous pteridophytes. Each whorl of leaves wraps itself around the stem at the point where it is jointed at the node. Sometimes the stems may branch, in which case these are also produced in whorls, which further enhances the already geometric form of the plant. The stems are marked by a series of longitudinal grooves and are hollow except at the nodes, where they are surprisingly tough and solid. Within the central cavity of the stem there is a ring of small cavities which alternate with the ribs of the stem. The size and shape of these cavities and the pattern of grooves on the stem are often important features in distinguishing one species from another. Toothed sheaths covering the joints are also important in identification. The number of 'teeth' on the sheath usually corresponds to the number of cavities in the outer ring of the stem.

In many species, including the field horsetail, there are two quite distinct types of stem. Fertile stems, which lack chlorophyll and are therefore purely reproductive in function, appear early in spring, usually in March, and once the spores have been shed they wither and are replaced by green vegetative stems.

The reproductive stems are usually quite stout, pale brown in colour and unbranched, the cone being carried at the top. Once more the whorled patterning is apparent in the cone, where T-shaped structures, called sporangiophores, twist around the central axis which is a simple extension of the stem itself. Each sporangiophore has a short cylindrical stalk (the pedicel) which expands at the top to produce a flat plate. The sporangia, which number between 5 and 10, hang beneath this plate and so many sporangiophores are present that they tend to become pressed together and are then hexagonal in shape. The spores are spherical and wrapped around them are two elastic band-like structures called elaters. When wet the elaters lie idle, but as they dry they become twisted and taut and eventually snap, catapulting the spore some distance from the parent plant. Whilst the spores are in the air, the elaters spread sideways to function as wings and extend the length of the spore's journey quite significantly.

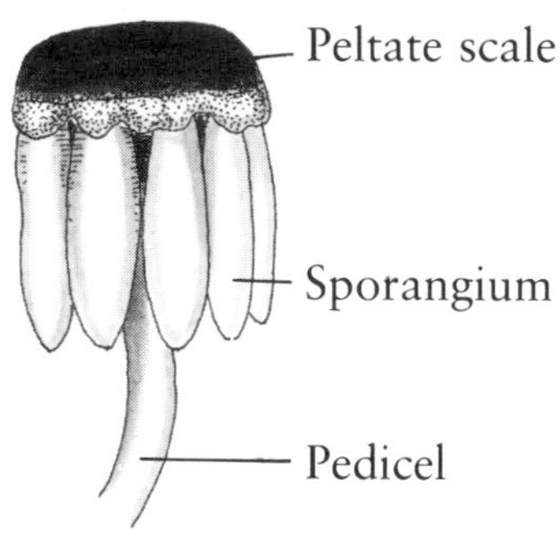

Equisetum arvense sporangiophore.

The spores of horsetails are short-lived and delicate compared to those of ferns, and are probably more suited to the warm, wet climate of the Carboniferous era. They do not cope well in the conditions of the present day. Indeed, the alternation of generations is probably less important to the success of Equisetum species than their spread by vegetative means. Nevertheless, there are conditions in which the cushion-like gametophytes do develop. The sexes are usually carried on separate prothalli and even if they do occur together, one stem matures before the other. The archegonium has a neck-like structure filled with mucus towards which the mobile antherozoid swims, propelled by a mass of flagellae situated at its apex. Soon after fertilisation, a root forms and pushes through the tissues of the prothallus, but this only functions until the stem has increased in length and developed its system of adventitious roots at the nodes. The stem initially grows upwards and only when there are around a dozen nodes does the creeping habit develop, by which time the horsetail sporophyte is safely established.

The alternation of generations in Equisetum follows the basic Pteridophyte pattern, but with a few fascinating variations on the theme. In contrast to the ferns there are two types of prothalli, one male the other female. Both are, however, detached from their spores in contrast to the condition seen in Selaginella. But the story does not end there. Although horsetail spores initially appear to be all of the same size, there is an apparently gradual variation from small to large. Botanists have discovered that there might be a small group and a larger group. If this is the case, then we might have the beginnings of a heterosporous situation. The ferns are without doubt homosporous, Selaginella is heterosporous, and Equisetum may be at an intermediate stage between the two, an interesting point from an evolutionary perspective.

Equisetum arvense, the field or common horsetail is, as we have seen, common, and an agricultural and horticultural pest. In the old days it was also called cornfield horsetail and children tired at harvest time, or awaiting their parents at the side of the field, staved off boredom by giving the stem a sudden pull. This separated it into small portions as the stems fractured at the node, leaving the sheath in which each was enclosed open to view. The number of 'teeth' on each light brown sheath varies from 6 to 12. The underground rhizomes bear tubers elliptical in shape and borne either singly or in strings. The species is found throughout the Arctic and north temperate regions. In Britain there are few habitats or counties closed to it, and several varieties have been suggested. However, none of these is sufficiently different to be regarded as a subspecies, and they are probably no more than variations in response to differing environments.

Great Horsetail

Equisetum telmateja, which we now term the great horsetail, has also been called the great mud horsetail and great water horsetail. Despite this latter name, the species is seldom found growing in water. The word 'telmateja' comes from the Greek meaning 'growing in mud'. The species occurs throughout Europe and also in the Mediterranean and north-western America. In Britain it is widespread, but not prolific, apart from in the Isle of Wight and parts of Northern Ireland. It is found in damp woods, sea cliffs and banks, but especially in areas where springs are bubbling from the ground.

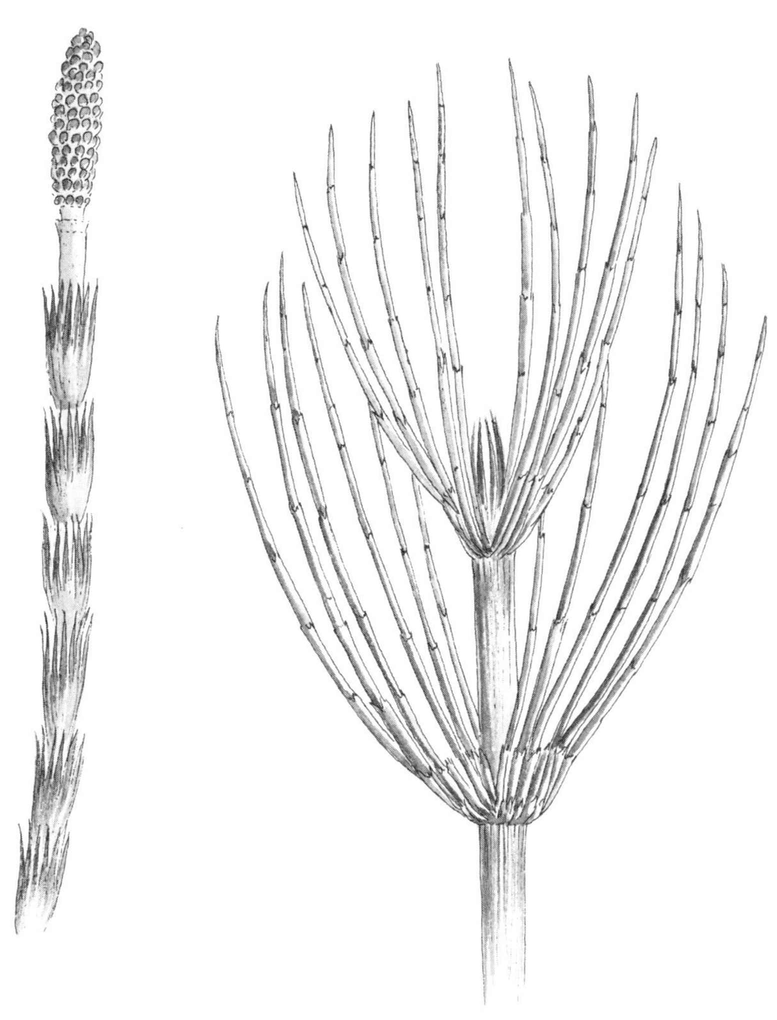

Great horsetail.

Equisetum telmateja is a tall, graceful plant and the largest of the British horsetails. The pale and barren stem which itself lacks chlorophyll is strikingly beautiful, often reaching heights in excess of 2m (6ft 6in), with whorls of delicate green branches giving the appearance of an oriental palm grove. Its symmetry, which has to be admired, is a result of its numerous upper whorls and branches 15 to 20cm (6 to 8 in) long. Towards the base the whorls are further apart and the branches are much shorter. The pale green stems are at their largest about the diameter of a stout walking stick, tapering gradually and becoming quite slender at the top. Their smooth surface is delicately marked with numerous lines, which become more distinct as they run into the sheaths. These structures are just over 1cm (0.4in) long. They clasp the stem closely and are green at the lower end, the upper portion being encircled by the dark brown ring. The 'teeth' are slender, dark brown with white edges and usually grow in groups of two or three. The branches frequently have, at their second joints, two to five secondary branches, each of which terminates in four or five teeth. Each of these extends into a slender black bristle with two toothed ribs. All of these factors are useful in identifying the species.

The fertile stems are much less significant, although they can reach 30cm (12in) in height, and are very easily identified. They are reddish-white, smooth, unbranched and with a distinctly succulent appearance. The upper sheaths are larger than the lower ones but all are quite loose and obviously funnel shaped. The sheaths are green at the base and dark brown at the upper end, and clearly marked with a number of lines. They have as many as thirty or forty teeth, which are long and slender. The strobili can be up to 10cm (4in) and each is basically conical, but narrowing to a blunt tip and wrapped around with whorls of sporangiophores. A very unusual feature of this species is the occasional appearance of a strobilus at the end of a stem which looks very like a sterile growth. The tendency to try to classify such plants as a subspecies or a variety has, fortunately, been resisted.

Wood Horsetail

Equisetum sylvaticum, the wood horsetail, is a rather pretty species which can be distinguished from other species even with only a casual glance. The pale green fronds are very reminiscent of a tiny Indian palm tree. It is found over most of Europe, temperate Asia and North America. The species is far more common in northern Britain than in the south and, despite its name of wood horsetail, it thrives on hillsides up to altitudes of over 609m (2,000ft). In Scotland some magnificent spreads of wood horsetail have been described, one of the most evocative descriptions being written by the Victorian botanist Newman.

'I observed it growing with peculiar luxuriance in the vicinity of Loch Tyne, in a little fir wood on a hillside. The fructification had entirely disappeared, and each stem had attained its full development, and every pendulous branch its full length and elegance. Altogether I could have fancied it a magic scene, created by fairies for their especial use and pleasure. It was a forest in miniature, and a forest of surpassing beauty.'

Despite this tendency to occur in the

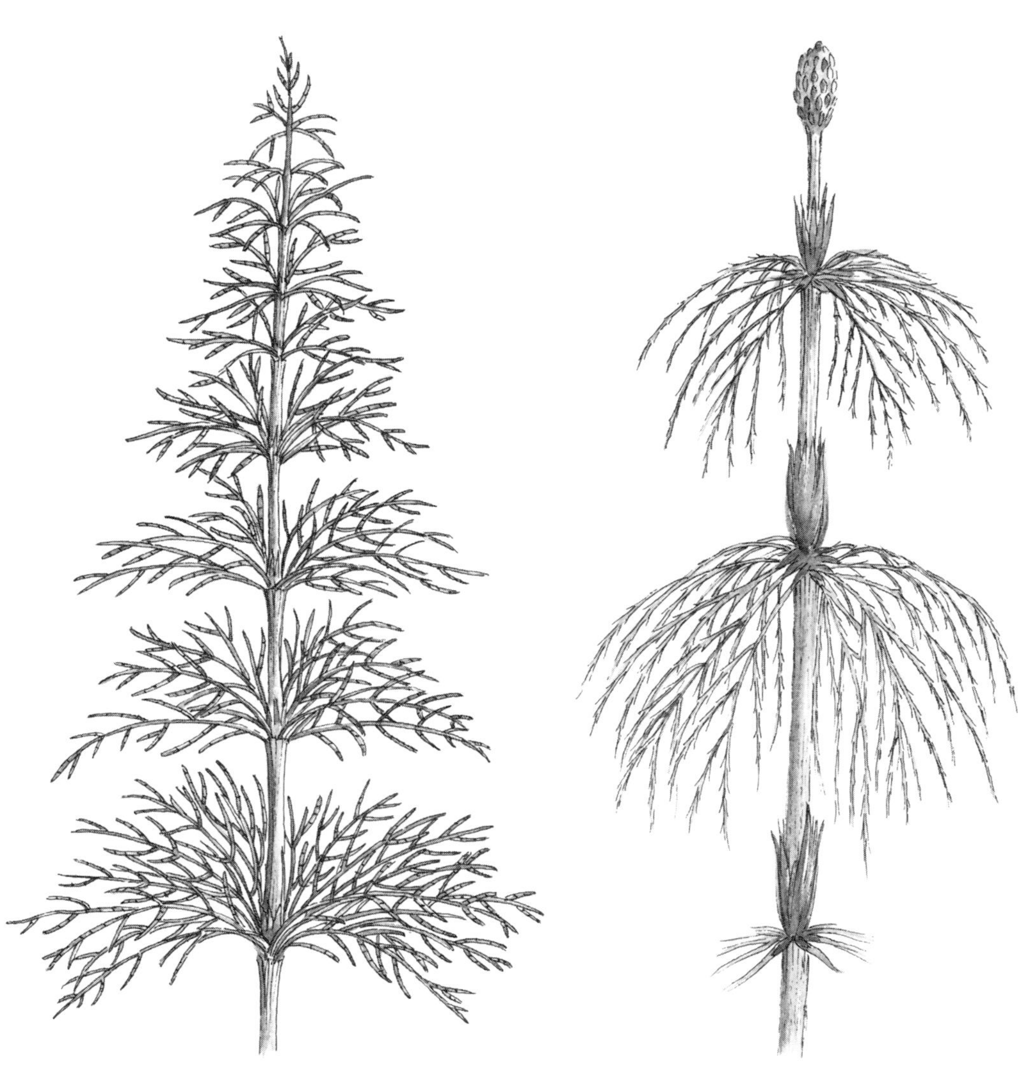

Wood horsetail.

north, in the south the species has been recorded in the Hampstead area since observed by Lobel in 1655.

The creeping rhizome is a dull brown in colour, covered with reddish-brown hair-like structures when close to the surface. At deeper levels the rhizome is naked, but branching is obvious, as are the tufts of fibrous adventitious roots. The tubers are egg shaped, around 0.6cm (0.25in) long, and are either found singly or in strings. Both the fertile and non-fertile stems bear leaves. The fertile stems are initially branchless, but these soon develop and vary in number from six to eight, growing from the nodes. The succulent stem is between 15 and 60cm (6 to 24in), is dull green and has about 12 slender ridges with corresponding furrows. It is not so rough or so firm as other horsetails. The margins of the large, loose sheaths are divisible into three or four lobes which are bright green at the base, the upper parts being brown. The sheaths have the same number of ribs as the stem. The whorled slender branches are around 5cm (2in) long and curve downwards. The diagnostic feature of this species is that these main branches have smaller branches between 1.25 and 2.5cm (0.5 to 1.0in) growing at the joints. In April the 2.5cm (1.0in) long tapering strobilus is pale brown in colour and perched atop a slender stalk. The mature spores are green in colour but are not seen very often, the species usually relying upon the vegetative propagation of the tubers on the rhizome for perpetuating itself.

The barren stems are much less succulent, much more slender and are far more liberally supplied with branches. The sheaths, although similar in shape, are smaller, more obviously ribbed and fit more closely against the stem. The compound branches, although crowded on to the stem, do not seem untidy, each measuring around 10cm (4in), and from every joint another whorl of branches of about half the length arises. These in turn may branch again and again, giving the whole plant the elegance of cascading greenery descending from the pointed tip. No horsetail can match it or be confused with it.

Shade Horsetail

Equisetum pratense, the shade horsetail, is found in Europe, Asia and North America, but is a rare plant in Britain, being confined to northern England, to the east of Scotland and a few sites in Northern Ireland. The specific name of 'pratense', meaning a meadow, is confusing to a British naturalist as the plant is associated with grassy river banks.

The fertile and sterile stems both tend to appear in the spring and exist side by side until autumn. The rigid fertile stems can reach 15cm (6in) in height, although there are records of specimens twice this size and all are of a delicate green colour which contrasts pleasantly with the loose sheaths, whose numbers vary from five to ten. Each is around 1.25cm (0.5in) long and has a pale green base, with the upper portion being brown. The structure is surmounted by between ten and twenty teeth, each one long and pointed and characterised by a dark brown longitudinal central stripe. Although the stem is not clearly ribbed, careful examination will show that the number of ribs is equal to the number of teeth. The blunt pale brown strobilus, with its darker tip, is about 2.5cm (1.0in) long and is supported on a short brown stalk.

The more slender sterile stems are pale green in colour and much taller, often reaching a height of 60cm (24in). They have a very rough texture, due to a covering of stout prickles, and some twenty ridges which are sharp enough to cut the careless hand which tries to pluck them. A few joints at the base lack branches but those higher up the stem put out whorls of between ten and sixteen drooping branches which gradually spread, forming larger circles. The rhizome is also very rough,

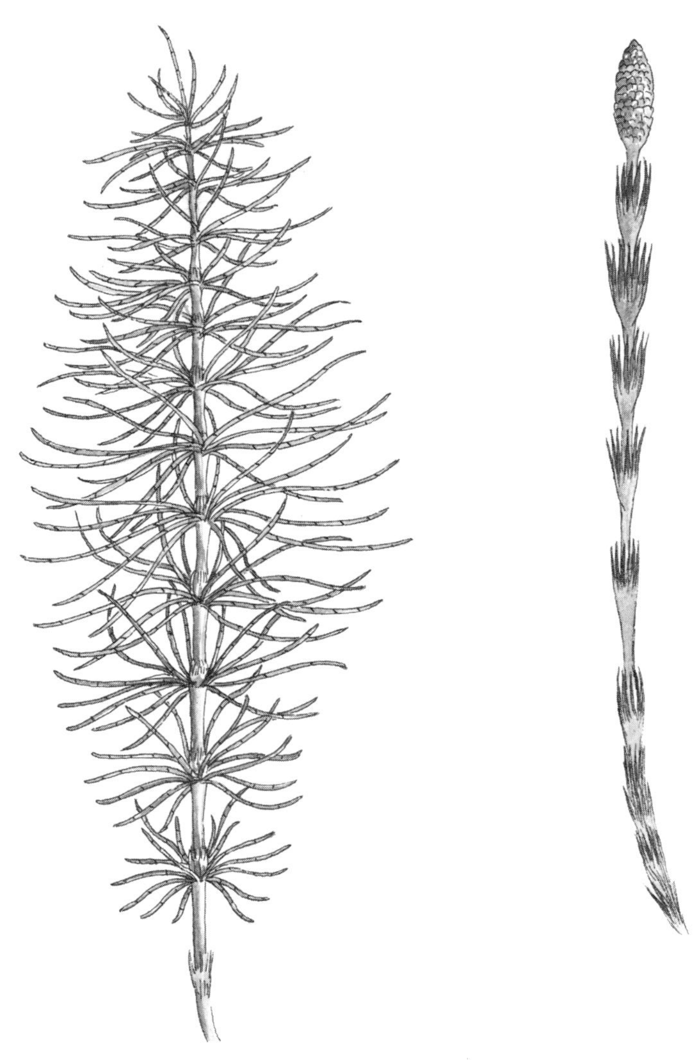

Shade horsetail.

especially in the regions situated above or closer to the ground, and this species appears never to develop tubers.

Marsh Horsetail

Equisetum palustre, the marsh horsetail, was well known to the Northamptonshire countryside poet John Clare (1793–1864) who wrote:

'Here Horse-tail round the water's edge
In bushy tufts is spread,
With rush and cutting leaves of sedge
That children learn to dread;
Its leaves like razors, mingling there,
All make the youngsters turn,
Leaving his rushes in despair,
A wounded hand to mourn.'

The long smooth and slender rhizomes, which are glossy and purplish-black in colour, can descend to depths of almost 1.8m (6ft). Black fibrous roots arise at the bases of the rhizome sheaths, and tubers are often present. They are glabrous (smooth) and ovoid but are arranged end to end rather like a string of sausages. As Clare's poem suggests, this is a common species growing among the wild flowers of the bog, a habitat suggested by its name 'palustre' meaning a marsh. It is widespread throughout the northern temperate regions of Europe as well as in Asia, including the USSR. It is also common in North America. Its habitats may vary from the damp slacks between sand dunes to the margins of upland tarns.

The fertile and sterile stems appear together in the spring and do not differ apart from the presence of the terminal cone (strobilus) in the fertile form. They also die down together in the autumn, living on the food-reserves made in the summer by the leaves and stored in the rhizome and especially in its associated tubers. The stems are invariably erect, about 40cm (16in) high and have about eight prominent, and as Clare noted, sharp ridges with deep furrows between them. The whorled arrangement of the leaves typical of horsetails occurs in *Equisetum palustre* but there are fewer branches towards the base, a good diagnostic feature. The joints are invested with large cylindrical sheaths which are pale yellow with darker tips and which loosely clasp the stem. The sheaths towards the upper end of the stem are almost twice as broad as the stem itself. The number of marginal teeth on the sheath is the same as the number of ribs on the stem.

The slender cone is around 2.5cm (1.0in) high and is carried on a stalk. Initially the whorls of scales are crowded into a black mass which separates, usually during May, to reveal white capsules attached to the margin. The structure is ripe during June at which time the strobilus (cone) is brown.

Most researchers have noted the variability of this species. An early writer on this subject was the Victorian naturalist Anne Pratt, who wrote:

'There are some singular varieties of this plant which, however, appear to be dependent on soil and situation, and not to become permanent. One form has been termed polystachion. Instead of the one cone usually placed, in the ordinary form of the Horse-tail, on the central stem, several of the branches of the two upper whorls terminate in cones, which are usually darker coloured than the commoner cone, more compact in form and appear later in the season. Another, and rarer variety, called nudum, is very much smaller than the ordinary plant, scarcely more than three or four inches high, having the lower part of the stem prostrate,

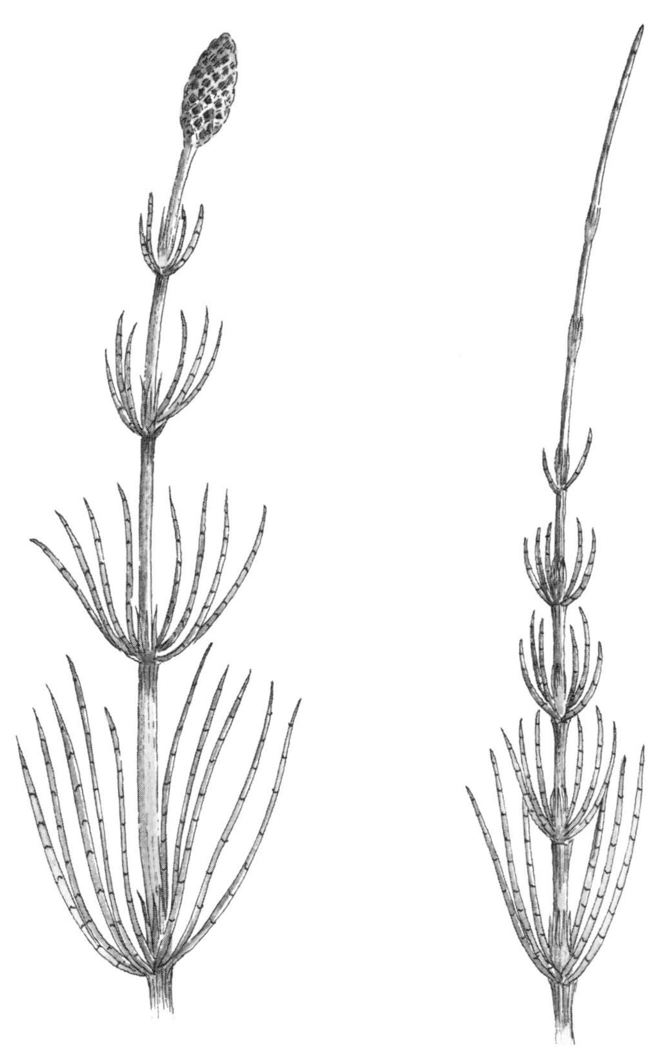

Marsh horsetail.

and the branches only about the base of its stem. It is apparently a dwarfed condition of the plant caused by want of nutriment. The form termed alpinum is very similar, and both are probably the result of growth on a soil less favourable to luxuriance or having been cropped by animals.'

Modern day botanists suggest that a type which they call 'var, polystachium vill' could be the result of damage to the apex of the main stem. This merely underlines

Anne Pratt's observation that horsetails do vary depending upon where they are growing. Field naturalists should therefore look for broad characteristics and not worry when a specimen differs a little from the norm.

Water Horsetail

Equisetum limosum, the smooth, naked or water horsetail, deserves the name 'limosum' which means 'muddy'. It is a pity, therefore, that many botanists now favour calling it *Equisetum fluviatile* indicating a tendency to grow near rivers. The species has a circumpolar distribution, being found throughout central and northern Europe, as well as in temperate Asia. In North America it stretches from Virginia northwards into Alaska.

Again we find that the water horsetail is very variable in structure, especially with regard to the degree of branching of the stems. These variations are found among the same population, often touching each other, and therefore there should be no question of a separation into subspecies or even varieties. In Britain the water horsetail is common and widespread in shallow water and on the fringes of rivers and lakes, and can occur both at sea level and around mountain tarns. The extremely smooth round stems and large cavities within the stem (which obviously give support in a damp habitat) can be seen with a botanical eyeglass and are diagnostic. The flinty texture typical of many horsetails is lacking because of these characteristics and this species is thus more useful to animals as a food, although it is reported that cattle in Britain will not eat it whilst it is green. Linnaeus, commenting upon the flora of his native Sweden in the 1750s, mentioned that tame reindeer preferred dried water horsetail to hay. Wild animals will also eat it, as recorded by the Victorian Knapp in his charming *Journal of a Naturalist*, describing the habits of the water vole *Arvicola terrestris* which he called the short-tailed water rat:

'A large stagnant piece of water in an inland county, with which I was intimately acquainted and which I very frequently visited for many years of my life, was one summer suddenly infested with an astonishing number of the short-tailed water rat, none of which previously existed there. Its vegetation was the common products of such places, excepting that the larger portion of it was densely covered with its usual crop, the smooth horse-tail. This constituted the food of these creatures, and the noise made by their champing it we could distinctly hear in the evening at many yards' distance.'

The long, slender, hairless and hollow rhizomes of this species are purple-black in colour and never bear tubers. Both the fertile and sterile stems put in an appearance in late spring, usually during May or early June, and persist well into the autumn. The stems are alike apart from the fact that the fertile stem bears an egg-shaped cone (strobilus), the sterile plant tending to taper to a point and being perhaps slightly taller. The usual height is between 60 and 120cm (2 to 4ft) although, as with all horsetails, a lot depends upon habitat.

Top: narrow buckler fern; centre right: hard fern; bottom: broad buckler fern.

Left: wood horsetail; centre: shade horsetail; right: field horsetail.

Top right: holly fern; bottom left: parsley fern; bottom right: limestone polypody.

Top left: hard shield fern; centre: black spleenwort; bottom: rustyback fern.

Top: green spleenwort; bottom left: moonwort; bottom right: adder's tongue.

Top: brittle bladder fern; centre: rigid buckler fern; bottom: soft shield.

Top: bracken; centre: mountain fern; bottom: common polypody.

Left: Wilson's filmy fern; top right: Tunbridge filmy fern; bottom right: Killarney fern.

Water/Field Horsetail Hybrid

At one time *Equisetum litorale* was listed as a discrete species but it is now thought to be a hybrid between the water horsetail *Equisetum fluviatile* and the field horsetail *Equisetum arvense*. Despite the unlikely combination, in view of the widely differing habitats of the two interbreeding species, the hybrid is widely distributed. It is, however, variable, resembling the water horsetail in wet conditions and the field horsetail in drier habitats.

Variegated Horsetail

Equisetum variegatum, the variegated horsetail, is a typical mountain species occurring in Wales, the Lake District and Scotland, but it also occurs in some sand dune systems principally those along the west coast. It is really an arctic species, the form of which varies a great deal according to habitat, and when growing on a sand dune it can appear to be a completely different species from those growing by mountain streams. The sand dune form seldom grows higher than 2.5cm (1.0in) and the stems are usually prostrate and consequently concealed among other dune plants, especially meadowsweet *Filipendula ulmaria* and water mint *Mentha aquatica*.

The species when growing at high altitude is more erect and can reach heights of around 90cm (3ft). The long slender rhizomes are almost black in colour and seldom penetrate very deeply into the soil, although the wiry roots which grow from the nodes provide sufficient anchorage. The slim unbranched stems grow during the summer but tend to persist for several years, the only difference between the fertile and sterile shoots being the presence of strobili on the former. Both tend to arise as groups close to the tip of the rhizome, giving the whole plant a tufted appearance.

The stems are rough to the touch, dark green in colour and have six to eight prominent ribs, about the same number as those found on the rhizome. Some of the stems can be erect, but others are prostrate and there may also be all possible permutations in between, so the variegated horsetail can be a difficult species to describe with precision. The loose, short sheaths are, however, a good diagnostic feature, even though they are only about 0.3cm (0.12in) in length. The lower part of the sheath is green, thus blending with the stem, and the upper portion is dark brown, contrasting with the white paper thin teeth. Each tooth ends in a point, has a dark line down its middle and the number of teeth on each sheath is the same as the number of ribs on the stem. In the fertile stem the sheath at the top of the stem is larger than usual and is folded around the black egg-shaped strobilus. This has a diameter of 0.6cm (0.24in) and is pointed at the tip.

Rough Horsetail

All the horsetails were at one time called 'polish grasses' and because of their rough texture, due to the particles of silica in their tissues, they were doubtless efficient pan scrubbers. Indeed, my great grandmother often used horsetails to assist with the washing-up following one of her magnificent picnics. Anne Pratt, writing in the 1880s, had this to say about the horsetails which were:

'. . . the plants sold in Queen Elizabeth's time by the "Herbe-women of Chepeside" under the names of Shave grass and Pewterwort, or Vitraria, though most would doubtless have been considered inferior to the *E. hyemale*, which Gerard calls "the small and naked Shave grass, wherewith fletchers (arrow makers) and comb makers doe rub and polish their work". It was very serviceable in the kitchens of olden times, and was doubtless used for cleaning the wooden spoons and platters; the "breen" of our forefathers as well as the "garnish" of pewter. Although in early days the tables of the opulent were served with silver, yet in humbler households wooden articles were commonly used at the daily meals until the fifteenth and sixteenth centuries, when pewter came into general use among the higher classes. However, it was not until the beginning of the eighteenth century that the articles made from it were sufficiently cheap to admit of their being seen at any save the rich man's table. Harrison, referring to this in 1580, says that in some places: ". . . beyond the sea, a garnish of good flat pewter of an ordinate making is esteemed almost so precious as the like number of vessels that are made of silver, and in manner no less desired amongst the great Estates, whose workmen are nothing so skilful in that trade as ours." The prices which he gives of the various articles prove their great costliness. The shave grasses served for cleaning either kind of ware, and this Corn Horsetail (common horsetail) is still used by dairymaids in Yorkshire for cleansing wooden milkpails; while the larger and less frequent plant, the Rough Horsetail, has long been known to our polishers of marble, and, under the name of Dutch Rush has been imported in large quantities from Holland for their use.'

This last sentence indicates that in many areas of Britain *Equisetum hyemale*, the rough horsetail or Dutch rush is not common, although it is widespread in its distribution and in some areas particularly in Northern England and Scotland the stands can be impressive. Drainage of many marginal sites has certainly reduced the species, but providing there is water close by, the rough horsetail is likely to occur. Shady banks overlooking streams and damp areas of woods are both ideal sites. Apart from Britain, and obviously Holland, it occurs throughout Europe and even as far south as the Mediterranean, as well as in the northern and central areas of Asia. It also occurs in some western areas of North America.

There are several distinguishing features which together render it unlikely that this species will be confused with any other horsetail. The deeply running and stout rhizomes can be up to 6m (20ft) and they always lack both tubers and hairs. They do, however, feel quite rough, due to a number of shallow ridges of which there are usually between ten and twenty-five. At the nodes, long slender roots with many branches arise, and these are usually covered with dark hair-like structures.

Both the fertile and sterile stems are unbranched, erect and only differ when the former is bearing its small black cone. During the summer they can reach 120cm (4ft) but are then quite likely to persist for several years. The specific name 'hyemale' means 'of winter' and indicates that stems can be found at this time. The plant contains so much silica in its tissues – up to 70 per cent – that it is very unlikely to rot quickly, although its colour changes from bright green when young to dull grey when past its best. Although tall, the stems are quite narrow, seldom exceeding 0.6cm (0.24in) in diameter, and they are covered with up to 30 longitudinal ribs, each consisting of a double line of rough points made almost entirely of silica. Small wonder that here was the medieval equivalent of sandpaper. Modern materials have now rendered the rough horsetail surplus to requirements, but we must never lose sight of these natural materials. They could still serve us well in an emergency and I still use them to scrub out the cooking vessels after a picnic.

The last three horsetails found in Britain are all rare and are unlikely to be encountered by the beginner. They are *Equisetum trachyodon* (the specific name means a 'tough tooth'), *Equisetum ramosissimum* (meaning 'many branched' horsetail) and finally Moore's horsetail, *Equisetum moorei*, named after Dr David Moore, a Scottish botanist. There are no more than half a dozen records in total, but the reader who wishes to complete the study of horsetails should refer to the *Atlas of Ferns of the British Isles* published jointly by the Botanical Society of the British Isles and the British Pteridological Society. This work also maps the distribution of hybrid types which are beyond the aims of this book. It provides an up-to-date appraisal of hybridisation and variations of British ferns which are useful once the basic principles of identification provided in this book have been grasped.

4 Ferns: Life History and Classification

Botanists describe the ferns, Megaphyllous Pteridophytes, as 'the vascular cryptograms' which means that their vegetative structure always shows the presence not only of strengthening tissue, but also of xylem, which transports water, and phloem, which transports sugar and salts. Cryptogram, as we have already seen, means that the reproductive processes were once considered hidden by comparison with those of flowering plants. Alternation of generations is clearly marked, but in contrast to the bryophytes, the dominant phase is the sporophyte and not the gametophyte. Furthermore, in the ferns the two generations lead totally separate existences. Up to this point we have described observable differences, but what is happening within the cells of the organism?

The cells of all living organisms, above the level of bacteria at least, are made up of a jelly-like cytoplasm, within which the chemical processes proceed. These are controlled from within the nucleus. A high powered microscope will reveal a varying number of tangled threads which stain easily with dyes and have therefore been named chromosomes (*chromos* meaning coloured, *soma* meaning body). Even higher power has revealed the chromosomes to be made up of genes, each of these complex structures controlling some characteristic of the species.

The reason why all human beings have similarities, and are recognised as a separate species, is due to the presence in our cells of 46 chromosomes called a diploid number. These are arranged in 23 pairs. Imagine what could happen when reproduction occurs – the next generation formed from a male and female cell could have 92 chromosomes, and four or five generations later the species would be extinct or changed. The solution to the problem is to reduce the number of chromosomes by half so that the offspring receives half of its genetic material from the father and the other half from the mother. Thus the human sperm contains 23 chromosomes and the egg also contains 23. This is called the haploid number. This explains why human babies are recognised as the same species as the parents and have similarities to both but are identical to neither.

In this context, the function of the alternation of generations found in the plants described here can be seen. The gametophyte generation reduces the chromosome count by half and when the gametes join together, the diploid number of chromosomes is restored to the sporophyte. We will look at how this scheme works in ferns, but first a good knowledge of fern anatomy is required. The best fern to examine is the very common male fern *Dryopteris felix-mas* which is the example chosen in biological text books.

The Male Fern

The name *Dryopteris felix-mas* has for many years been an all-embracing word for a complex of species rather than a single type. Much painstaking work has been done involving the counting of chromosomes within the cells by several eminent botanists including C.R. Fraser-Jenkins. There is still some confusion but *Dryopteris felix-mas* seems to be a woodland-based species favouring light soils. *Dryopteris oreades* prefers open hills and screes whilst *Dryopteris pseudomas*, although also a woodland species, prefers clay soils and maritime conditions tending to retain its colour throughout the winter. The three species differ in the structure of the rhizome, shape of fronds and, of course, chromosome number. The lesson to be learned here is a vital one. Anyone working on ferns should keep abreast of any new developments and some help has been given in the back of this present book in the form of a list of societies and journals. The basic internal anatomy of the above complex of species is, however, the same, as are the main events in the alternation of generations.

External Morphology

Dryopteris felix-mas has a rather stubby, usually unbranched rhizome growing obliquely in the soil. A mass of tough fibrous adventitious roots sprout from the rhizome (underground stem) and penetrate the soil in all directions, providing a very firm anchorage. Because of its oblique positioning, the apex of the rhizome approaches close to the soil level and from it arise a cluster of large leaf-like fronds. The rhizomes themselves vary a great deal in size and since they function as food storage organs during the winter, the older the rhizome the larger it tends to be. The older parts, more distant from the apex, will eventually wither and die. Indeed, a substantial proportion of the underground structure consists of the dead and blackened bases of dead leaves which tend to remain long after the upper portions have broken away and decayed. The whole rhizome looks very solid due to a covering of structures called ramenta which are dry brown scaly outgrowths and which are wrapped around the bases of the leaves.

The leaves, or fronds as they are called, are described as compound which means that they are made up of a number of distinct parts. Each frond consists of a main axis called the rachis, bearing leaflets called pinnae which are further divided into pinnules. The swollen base of the rachis is called the petiole and this merges with the rhizome at one end and tapers to the end of the frond at the other. Each frond takes two years to develop fully and the mature fronds are obviously those furthest from the apex. They can be as much as 1.0m (3ft 3in) in length. Huge fronds like this are very different from the tiny structures found in the club-mosses and horsetails, accounting for the ferns being classed as megaphyllous pteridophytes.

The very young leaves are folded in what is described as circinnate fashion, which means they look rather like a bishop's crosier, with the rachis, the midribs of the pinnae and pinnules coiled in a spiral-like manner. The order of unfolding from leaf buds is described as its vernation, and the method described above is a primitive characteristic employed by *Dryopteris felix-mas*. Other species, as we shall see in later chapters, may have different methods of vernation.

The very young leaves of the male fern are protected by a prolific covering of ramenta. Fully developed leaves eventually bear specialised reproductive structures on the lower (abaxial) surface. These are called sori and within them develop the sporangia. Once a frond is mature enough to bear sporangia it is referred to as the sporophyll. In time each and every frond develops sori and, unlike some fern species, no specialised reproductive leaves are developed.

The sori lie across the forks of the veins on the sporophyll and seven or eight may be present on each pinna. The total reproductive capacity of one fern is therefore staggeringly large. During the early months of the summer each sorus is covered and protected by a green kidney-shaped scale called the indusium. As the season advances, the indusium withers to reveal dark brown maturing sporangia and if the frond is shaken at this time, a cloud of 'dust' hangs in the air as the spores are liberated.

The sporangia do not all develop at once. Whilst one crop is releasing spores, another group is maturing to ensure a release of ripe spores over a period which is more likely to result in successful germination. Each sporangium is biconvex in outline and is carried on top of a slender stalk. The cell wall is only one cell thick and within this are the tapetal cells, which themselves enclose a dense material called the archesporium. This material divides to produce sixteen central cells which eventually separate from each other and lie in a fluid-filled cavity produced by the breakdown of the tapetum. For obvious reasons the cells are called spore-mother cells, and each undergoes two divisions to produce four spores which are initially grouped tetrahedrally in fours.

Whilst these divisions are proceeding, the wall of the sporangium also changes and a structure called the annulus develops, consisting of a single line of cells along the narrow margin. This extends from the stalk on one side, over the top of the sporangium, and about half-way down the opposite side. The net effect of the development of the annulus is to produce a sporangium of uneven thickness, and as it dries the water-loaded cells of the annulus are pulled inwards, causing a rupture in the sporangium. The work of the annulus is still not completed, however, because the strip of cells straightens out under pressure and finally snaps back into its original position like an elastic band, hurling the spores out from the sporangium as it does so. The spores are tiny, and the dispersal mechanism ensures that they are released in dry weather. In the slightest of breezes, they can often travel considerable distances.

Some spores, but possibly only a small proportion of those produced, land in moist conditions where they germinate to produce a flat heart-shaped green plant known as the prothallus. This is the gametophyte generation of the life cycle, and it seldom measures more than 1cm (0.4in) across. It is a flattened structure which botanists describe as 'dorsi-ventral' and from its under-surface arise unicellular rhizoids. These are developed from a central cushion-like structure which is several cells thick. They serve to anchor the prothallus which never develops any of the vascular tissue so typical of the sporophyte generation.

The gametophyte is, however, a completely independent structure which is capable of producing its own food by photosynthesis and can absorb both water and mineral salts through its rhizoids. It is none the less a very delicate structure and not well adapted to live in dry conditions. It has no protective cuticle covering its surface, so it can only survive in the dampest of habitats, where the risk of desiccation does not arise. Its main function, however, is to produce the gametes as soon as possible, and a long life is not part of the plan. The male organs (antheridia) appear rather early in the development of the prothallus and are widely scattered among the rhizoids. The female organs (archegonia) are found at a later stage and are situated on the 'cushion' just behind the apical notch of the prothallus. This phased development of the sex cells ensures that the gametophyte does not usually fertilise itself, and thus guarantees a regular mixing of the genes which is obviously good for the species.

Each antheridium arises as a single superficial cell on the lower side of the prothallus. Early divisions of this cell isolate a central colourless cell from a layer of wall cells which contain chlorophyll. Eventually each antheridium develops a short stalk and therefore resembles a round protuberance on the prothallus. Within the protective wall the central cell then begins a period of rapid division to produce 32 antherozoids, each containing half the normal chromosome number. Each antherozoid has a group of long flagellae which whip the animal through a film of water towards the female organ.

Each archegonium also arises as a single superficial cell on the lower side of the prothallus. A series of divisions produces a typically cryptogramic female organ consisting of a lower venter, embedded in the tissues of the prothallus, whilst a curved neck actually projects from the surface. This provides the antherozoid with a point of entrance and probably also provides a chemical stimulus to attract it. The chemical has been identified as malic acid. The venter contains the female oosphere (egg) and below this is the central canal cell. The neck consists of four rows of cells which are joined laterally around a central canal, the curvature being due to the fact that there are five or six cells on one side and only four on the other. The immature archegonium is characterised by the presence of neck canal cells which effectively seal the entrance. Eventually the ventral canal cell and the neck cells break down and the tip ruptures, leaving a passage full of mucilage right up to the venter through which the antherozoid can swim and fuse with the oosphere. After being fertilised, the oosphere becomes known as the zygote, which then secretes a wall around itself and is then called the oospore. Usually several oospores are formed in each gametophyte prothallus but only one of these gives rise to another sporophyte generation.

Although the oospore does have a resting period it germinates soon after the venter of the archegonium has decayed and thus released it. A series of cell divisions then produces eight cells. Four of these grow towards the apex of the parent gametophyte, producing the main, and the first, stem, whilst the other four cells grow away from the apex. These produce the primary root and an absorptive organ called the foot, which remains embedded within the prothallus and draws nourishment from it. Until the frond begins to produce food, the sporophyte generation is actually living parasitically upon the gametophyte. As the sporophyte becomes self-sufficient, the gametophyte tissues which still remain wither away and die.

Compared to bryophytes and other pteridophytes the spermatophyte of the male fern is a somewhat curious combination of highly evolved and primitive characteristics. In the vegetative sense, Dryopteris is highly developed, having a distinct root, stem and leaf system, and it is also an efficient perennator. There is a great deal of tissue specialisation with specialised cells photosynthesising, and others, called parenchyma, storing the material produced. If required, the food and salts can be moved around the plant in the phloem made up of sieve cells, whilst water is moved through xylem cells called tracheids. This is a great advance on the situation found in the bryophytes, which also lack the mechanical support tissues common in the ferns.

The pteridophytes, however, lack certain features which typify the flowering plants. In the phloem, the sieve cells do not combine to produce larger and more efficient sieve tubes, and there are no xylem vessels which have a greater capacity than the tracheids. Growth in ferns is from a single apical cell and thus neither the complex branching of higher plants, nor the formation of wood typical of woody perennials is possible. In contrast with the sophisticated specialisation shown by the sporophyte, the gametophyte is less well adapted to life on land than the majority of mosses.

With regard to reproductive strategies the same trends are evident, parts of the process being highly advanced whilst others – obviously the reason why ferns have not proved to be the dominant plant group today – are primitive. The asexual sporophyte copes easily with dry conditions and actually relies on fine weather for the release of the spores. The sexual gametophyte is completely dependent upon wet conditions and neither the structure of the sex organs, nor of the mobile male cells (zoidogametes) shows any advance on the condition found in the bryophytes.

By comparison with other pteridophytes such as the club-mosses and horsetails, the ferns also show some advanced features but other features are less well developed. The macrophyllous condition and the advanced anatomy both place the ferns ahead of the other pteridophytes in the evolutionary sense, but with regard to reproduction the ferns are behind the others. Within the pteridophytes there is an increasing tendency for the sporangia to be borne on special fronds called sporophylls and for these to be aggregated into compact structures called cones or strobili which we have already described at length in Chapter 3. Although some ferns do have separate reproductive fronds, the absence of strobili is regarded as a primitive condition.

Two other evolutionary trends include homospory and apospory. Homospory means that all the asexual spores are identical and contrasts with heterospory in which there are two types of spore, each of which produces two distinct types of gametophyte, as seen, for example, in the club-moss Selaginella. The heterosporous condition certainly evolved from the homosporous and is regarded as one of the earliest steps towards the development of the seed-bearing higher plants. Apospory is the term used to describe the development of a gametophyte directly from the sporophyte without the production of spores. Apogamy is the word used to describe the development of a sporophyte directly from the gametophyte without the union of gametes. Both conditions occur regularly among ferns and a single species may occasionally demonstrate both apospory and apogamy. The net result is to shorten the life cycle, and perhaps also to provide some degree of independence from water during reproduction.

Classification

There are two ways in which the British ferns could be described. Firstly they could be described family by family in the form of a taxonomic list. This would make sense if this book was aimed at the pteridologist rather than the naturalist. There are, however, already books of this type, but these usually include all the European species as well as many rarities and subspecies. The second approach is to describe the ferns found in particular habitats so that the naturalist wanting to know which ferns grow where, and how to identify them, is satisfied. This is the purpose of this book, but first a brief mention of how British ferns are classified is essential.

We have around forty-six species of fern, some of which are quite rare, arranged in thirteen families. The species themselves are fairly easily defined, but experts in the field are always describing new hybrids, subspecies and varieties. Beginners should not become confused by this, but should concentrate all their available energy at the full species level.

Family 1 Ophioglossaceae

This family includes the moonwort *Botrychium lunaria* and adder's tongue *Ophioglossum vulgatum*. Both these species were probably once widespread but have decreased recently due to our demands for more land. They both still grow in upland areas and in dune slacks and are described in Chapter 6. Two further species should be mentioned, although neither is described in the main text, and neither has a vernacular name. *Ophioglossum lusitanicum*, which might be called the dwarf adder's tongue, is a species found mainly in the Mediterranean and its most northerly site is in the Isles of Scilly, but it also occurs in the Channel Islands. Some workers have suggested that it may also occur in the south-western tip of Ireland but could have been overlooked because it dies down for most of the year and is only usually visible between November and February. There are more records of *Ophioglossum azorcium* from northern Scotland to southern Britain – all the records being coastal, particularly on salt marshes. Its relationship with adder's tongue is, however, far from clear and some feel that it may not warrant the status of a full species.

Family 2 Osmundaceae

The Royal fern *Osmunda regalis* is the only British representative of the family and because of its preference for areas of high rainfall it is described in Chapter 5.

Family 3 Adiantaceae

There are three British species of which the upland based parsley fern *Cryptogramma crispa*, described in Chapter 8, is the most common, even though its range is somewhat restricted. The Jersey fern *Anogramma leptophylla* is restricted to the Channel

Islands as its name implies and is therefore included under the heading of sea based ferns in Chapter 6. Also described in this chapter is the rather rare maidenhair fern *Adiantum capillus veneris*.

Family 4 Hymenophyllaceae

Members of this family often bear a superficial resemblance to the liverworts and are usually found close to running or dripping water. The three British species of the mainly tropical family are therefore described in Chapter 5, which covers ferns of the waterside. Each has a restricted distribution. They are the Tunbridge filmy fern *Hymenophyllum tunbrigense*, Wilson's filmy fern *Hymenophyllum wilsonii* and the Killarney or bristle fern *Trichomanes speciosum*.

Family 5 Polypodiaceae

When dealing with the British members of this family we find ourselves faced with controversy. Many feel that we have only one representative, namely the common polypody *Polypodium vulgare*, which occurs on tree trunks, cliff sides and especially walls throughout Britain. It is therefore described in Chapter 9, a decision which may not please the purists, who want to suggest seven subspecies or hybrids, some of which bear no common names. These are the polypody *Polypodium vulgare agg.* which they feel may have resulted from the interbreeding of all, or some, of the others including the common polypody *Polypodium vulgare*. Then we have the intermediate polypody *Polypodium interjectum* and *Polypodium x mantoniae* which is a hybrid between the common and intermediate polypody. There is also the southern polypody, *Polypodium australe*, which is said to be a mainly Mediterranean species, but reaches its northernmost limit on the limestone island of Lismore off Oban in Argyll. Finally, there are two more hybrids to consider; *Polypodium x font-queri* is a cross between the southern and common polypody; *Polypodium x shivasiae* is an uncommon and sterile hybrid between the southern and intermediate polypody. Beginners should not concern themselves with the ever-changing taxonomy of the polypody 'agg.', but need to be aware of it as they read the literature.

Family 6 Dennstaedtiaceae

Despite the fact that we have only one species to represent the family in Britain, it certainly makes its presence felt. It was difficult to decide in which chapter to describe the ubiquitous bracken *Pteridium aquilinum*. In the end I decided to include it in Chapter 9 but it could have been included under any heading. Suggestions are already being made that the bracken should be separated into subspecies perhaps reflecting different habitats. Fortunately, however, bracken still remains uncomplicated by the activities of taxonomists.

Family 7 Thelypteridaceae

This family is represented in Britain by three species. The marsh fern *Thelypteris thelypteroides* is described in Chapter 5, the beech fern *Phegopteris connectilis* has a pronounced west and north distribution in Britain and is described under the heading of woodland ferns in Chapter 7. The lemon-scented fern *Oreopteris limbosperma* is, because of its alternative name of mountain fern, described in Chapter 8.

Family 8 Aspleniaceae

As with the polypodies we find some difficulty with the precise taxonomy of the spleenwort family because of the number of variations recognised by experts. Most field naturalists would be happy to accept nine species, and the present volume follows this trend. Hart's tongue fern *Asplenium scolopendrium* is widely distributed providing there is some lime present in the soil and it appears to have no subspecies, although in cultivation many varieties have been developed. It grows well in the crevices of walls, obviously thriving on the calcium rich mortar, and is therefore described in Chapter 9.

Hybrid types between hart's tongue and other spleenworts have been described but are unlikely to detain us very long, however expert we may be. *Asplenium x confluens* is a cross between hart's tongue and the maidenhair spleenwort *Asplenium trichomanes*, which has been recorded only at Levens park in Cumbria in 1865, and at Whitby and Killarney in 1875. This again underlines the need for amateur naturalists to stick to descriptions of species and to avoid the sometimes literal splitting of hairs. *Asplenium x microdon* is likewise a very rare cross between hart's tongue and the lanceolate spleenwort *Asplenium billotii.*

The black spleenwort, *Asplenium adiantum-nigrum*, described in Chapter 9, is a widespread and discrete species capable of withstanding an often high degree of exposure. Once more, however, we find a number of potentially confusing hybrids which are, thankfully, quite rare. *Asplenium x jacksonii* is a cross between the black and hart's tongue spleenworts, but the list of hybrids and potential hybrids is increasing apace. With regard to the serpentine black spleenwort we appear to be on safer ground and the species called *Asplenium cuneifolium* (the specific name describing the frond shape) is confined to the serpentine rocks of Scotland, Cornwall and the extreme west of Ireland. Serpentine is a type of asbestos based on a hydrated silicate of magnesium. The sea spleenwort *Asplenium marinum*, described in Chapter 6, is confined to the sea coast where it is struck by salt spray. There seem to be no problems or complications with regard to its life history. The same certainty does not apply to the maidenhair spleenwort described in Chapter 9 under the heading of *Asplenium trichomanes agg.* Taxonomists recognise many subspecies, the two most commonly described being *Asplenium trichomanes trichomanes* which is common on mountains, and *Asplenium trichomanes quadrivalens* which shows a preference for walls containing mortar, especially those in wetter parts of the country. Once more the differences can be quite subtle and the interested reader is referred to the bibliography. The green spleenwort *Asplenium viride* is almost certainly a discrete species found in the highlands of Britain (with a few exceptions) and it is therefore described in Chapter 8.

The wall rue *Asplenium ruta-muraria* which, as its name implies, is best described in Chapter 9, could be called a discrete species, although two possible hybrids have been noted. *Asplenium x murbeckii* is thought to be a cross (observed in the Lake District) between wall rue and the forked spleenwort *Asplenium septentrionale.* In the north-east of Ireland a cross between wall rue and *Asplenium trichomanes quadrivalens* has been suggested.

The forked spleenwort *Asplenium septentrionale* is another mountain based species, and fairly rare, doubtless due to its geological demands discussed in Chapter 8. A suggested cross between this species and *Asplenium trichomanes trichomanes*, the maidenhair spleenwort, has been described and named *Asplenium x alternifolium*. The final member of the British spleenworts, and described in Chapter 9, is the rustyback, *Asplenium ceterach*. This is another lime loving species and it is often found on walls which are mortared. As yet there seem to be neither subspecies nor hybrids with other spleenworts.

Family 9 Athyriaceae

This is another large family with representatives in a variety of habitats but by far the most widespread species is the lady fern *Athyrium felix femina*. This is most common in deciduous woodlands and is therefore described in Chapter 7. The alpine lady fern *Athyrium distentifolium* is an upland species described in Chapter 8, as is the even rarer *Athyrium flexile*, both species being confined to the Scottish Highlands.

Much more widespread, but also upland species, are the oak fern *Gymnocarpium dryopteris* and the limestone fern *Gymnocarpium robertianum*. The brittle bladder fern *Cystopteris fragilis* is also found in upland habitats but shows a preference for wet areas and is occasionally recorded in southern areas, particularly in the west. It is therefore described in Chapter 5.

Casual fern hunters are hardly likely to come across Dickies bladder fern *Cystopteris dickieana* which is only recorded from one seaside cave in Kincardine, Scotland. It is, however, thought to occur in some of the mountains of northern Britain, although I have briefly described it in Chapter 6. There are no doubts about placing the mountain bladder fern *Cystopteris montana*, the oblong woodsia *Woodsia ilvensis* or the Alpine woodsia *Woodsia alpina* in Chapter 8, since they are all true mountain species.

Family 10 Aspidiacea

This is a very large family, with perhaps thirteen species, although the taxonomy is complex, many hybrids having been described purely from herbarium specimens. Fortunately the identification of most of the species is quite straightforward. Without doubt the holly fern *Polystichum lonchitis* is an arctic Alpine species described in Chapter 8 and although a hybrid with the soft shield fern *Polystichum setiferum* has occurred, it was only recognised after detailed microscopic investigation. The soft shield fern prefers a warm, wet climate and is often found in the hedges of southern England. It is therefore described in Chapter 9. The hard shield fern also occurs in the same habitat, having proved surprisingly resistant to pollution. Indeed *Polystichum aculeatum* grows well alongside man-made structures such as roads, railways, canals and especially on bridges. Hybrids have been described between the holly fern and hard shield-fern and also between hard and soft shield-ferns, although they are recorded from herbarium specimens and are not likely to concern the ‘trainee’ fern hunter.

The mountain male fern *Dryopteris oreades* is often found on mountain scree and is therefore described in Chapter 8. It has been confused with both the scaly male fern *Dryopteris pseudomas* and *Dryopteris felix-mas*, the male fern, both of which are mainly woodland species and described in Chapter 7. The points of distinction will be made there but the field naturalist should never fail to take differing habitats into account during the often long process of precise identification. To say that the taxonomy of *Dryopteris felix-mas* is complicated at the moment is an understatement, but once armed with the pteridological language, the literature can be quite fascinating.

The hay-scented buckler fern *Dryopteris aemula* is a mountain based species over most of its range, but in Britain it has a definite oceanic distribution and it is therefore described in Chapter 6, whereas the rigid buckler fern *Dryopteris villarii* is a species restricted to limestone pavements and is therefore described in Chapter 8. Taxonomists are of the opinion that this is a complex species, the British plants being classified as *Dryopteris villarii submontana* and apparently having double the number of chromosomes as those found further to the south in Europe. The crested buckler fern *Dryopteris cristata*, described in Chapter 5, is a water loving species once common in the fens of Britain but now almost completely confined to the Norfolk Broads. Much more widespread and occurring in wet woodlands is the narrow buckler fern *Dryopteris carthusiana* which is described in Chapter 5, along with the broad buckler fern *Dryopteris austriaca*. The northern buckler-fern *Dryopteris expansa* is an arctic Alpine species which should therefore be correctly placed in Chapter 8, but it should be noted that it is frequently found in the woods of western Scotland right down to sea level.

Some hybridisation has also been noted among the Dryopterids. *Dryopteris x brathaica* has only been recorded once on the north-western bank of Lake Windermere and is a hybrid between *Dryopteris felix-mas*, the male fern, and *Dryopteris carthusiana*, the narrow buckler fern. The latter has also hybridised with *Dryopteris cristata*, the crested buckler fern. This is another rare hybrid, confined to Norfolk and called *Dryopteris x uliginosa*. *Dryopteris x deweveri* is, on the other hand, a common hybrid between the broad buckler fern *Dryopteris austriaca* and the narrow buckler fern *Dryopteris carthusiana*. There is also a commonly occurring hybrid whenever the northern buckler fern *Dryopteris expansa* is found in the same place as the broad buckler fern.

Family 11 Blechnaceae

Only the hard fern *Blechnum spicant* represents this family in Britain. It is widespread, apart from in limestone areas which do not appear to suit it. It could be described in several habitats but it has been included in Chapter 7 as it is so common in wet woods, whether dominated by deciduous or coniferous trees.

Family 12 Marsileaceae

The pillwort *Pilularia globulifera* is thinly spread throughout Britain but is losing considerable ground as wetland areas, and especially ponds, are drained. It is described in Chapter 5. This small family is typified by long hairy rhizomes carrying sporangia contained within organs called sporocarps, which are of two types and produce megaspores and microspores. The genus Pilularia is widespread and found in Australia, America and Europe. It includes six species but only one occurs in Britain.

Family 13 Azollaceae

The single British representative of this tropical and subtropical family, the water fern *Azolla filiculoides*, is also described in

Chapter 5. It seems to be spreading in Britain although it is thought to have been originally introduced from tropical America.

5 Ferns of the Waterside

In one sense the British climate with its high rainfall, especially in the west, and its moderate winter temperatures compared to those in Europe, should have favoured a prolific growth of ferns. Two factors, both involving human interference have worked against this. Firstly, the demand for good land led to the draining of marginal land along rivers, close to ponds, and vast areas of fenland, all ideal fern country. Add to this the Victorians' urge to collect ferns, especially the rare or the exotic looking, and one can see why some of our waterside ferns are now uncommon.

The Royal Fern

Human activities have made it difficult to plot the natural distribution of the royal fern, *Osmunda regalis*, with any certainty. It is such a beautiful plant that collectors have planted it throughout the country and spores escaping from the gardens have resulted in its appearance in the most unlikely places. There is no doubt that it prefers areas of high rainfall and boggy areas are thus favoured. This suggests the reason for its preference for western areas, even those blasted by sea breezes laden with salt. The most impressive growths of the royal fern that I have ever found in Britain have been on the Western Isles of Scotland, although it also grows well in the west of Ireland and in Cornwall. It is widely distributed throughout the world in both tropical and temperate regions.

No doubt its striking physical appearance has earned *Osmunda regalis* its vernacular names of royal and regal fern, but the old name of 'flowering fern' is very misleading. Perhaps it arises because the species produces fertile and sterile fronds but no cryptogram ever produces flowers. The generic name of Osmundia is interesting, and is thought to derive from 'osmund', a Saxon word meaning strength and encompassed in the names of Edmund and Sigismund. Osmundia was one of the names given to the Celtic god of thunder, Thor. Others have suggested that the word means domestic peace – 'os' meaning a house and 'mund' meaning strength.

In the days before the urban sprawl from London encompassed the surrounding villages, the royal fern must have been common. Gerard, the seventeenth-century herbalist, recorded it from Brentwood. He had obviously cut into the tissues of the plant and refers to the white centre as 'The heart of Osmund the waterman'. He was alluding to a tradition that a Loch Tyne waterman had fought off a group of invading Danes, and then escaped after hiding his family among the tall fronds of this plant. It also seems to have had other and more regularly used functions and was reported to be:

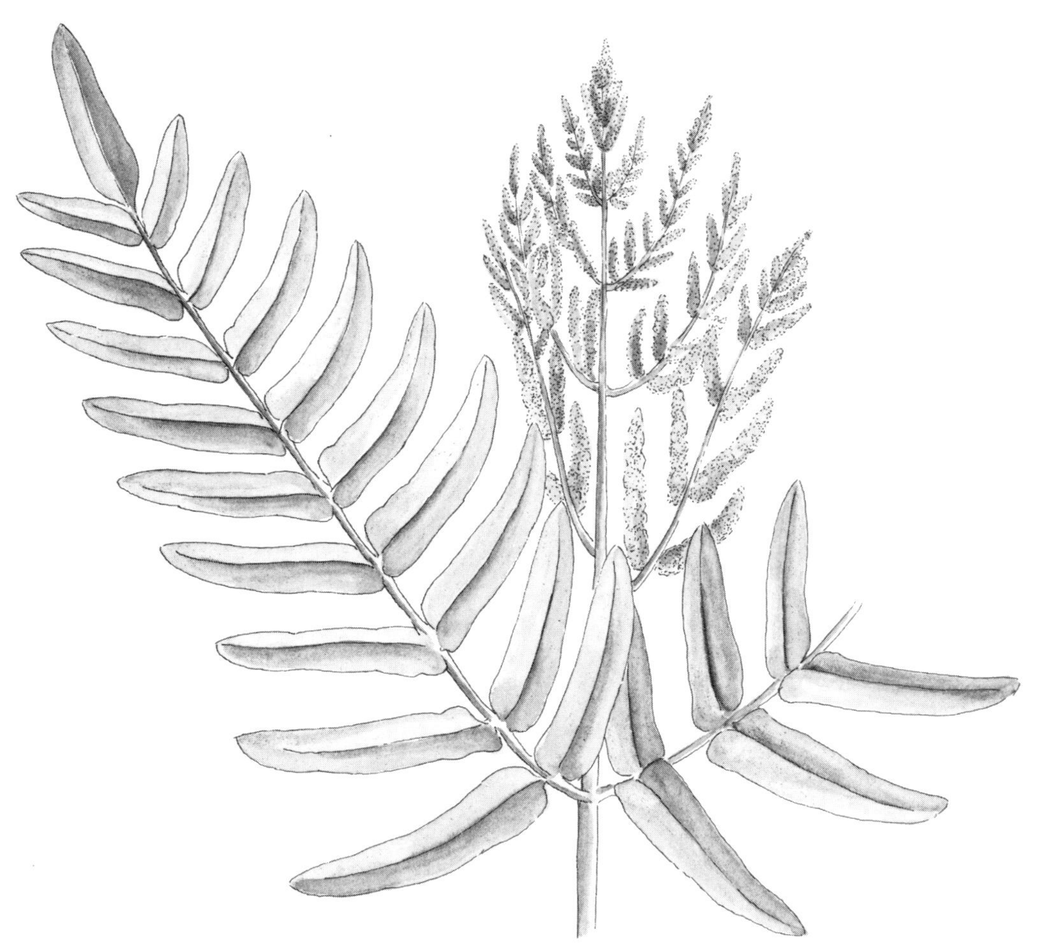

Royal fern.

'good in wounds, bruises and the like. The decoction to be drunk, or boiled into an ointment of oil, as a balsam or balm; and so it is singular good against bruises and bones broken or out of joint.'

The stem was used as both an astringent and a tonic. The root, when slowly simmered in water, breaks down to a mucilaginous mass which was in great demand as starch for stiffening linen. The species would therefore be likely to be well known to country folk throughout Europe and it was called *osmonde* in France, *osmunda* in both Spain and Italy, *traubenfarrn* in Germany and *trosvaren* in Holland. This is why we need the scientific name, and all nations recognise the royal fern as *Osmunda regalis*.

Identification

The plant grows from a stout perennial rhizome. The fronds, which are sensitive to frosts, do not usually appear until May and begin to die back during August. There are normally ten to twelve young fronds which initially are thin and crisp. At first the colour is pale green, but this deepens with age. The stalk (or stipe) of a young frond is an attractive shade of reddish brown but this also turns green in the mature frond. Royal fern fronds vary in height according to the habitat in which they are growing but they can reach 3.7m (12ft) and the whole plant looks surprisingly shrub-like.

The fronds are described as lanceolate (lance shaped and tapering towards the tip) and twice pinnate which means they are divided like a feather, with a central rachis and opposite rows of pinnae leading off it. The pinnae (leaflets) of the frond are themselves lanceolate and bear even smaller structures called pinnules which are ear-shaped at the base, rounded towards the apex, with serrated edges. The fertile and infertile fronds differ very little except that the pinnae towards the top of the former are densely covered with brown clusters of sori. These upper pinnae do look like spikes of flowers, especially from a distance, and so perhaps the old name of 'flowering fern' is not so wide of the mark as we thought. The fact that the sterile fronds tend to be towards the outside of the cluster and droop at their tips adds to the attractions of this delightful species.

Individual plants are thought to be able to live for several hundred years and have proved popular with gardeners but the root-stocks have another function. Many epiphytic orchids have been found to grow very well on the stout underground organ of the fern.

It is reported that Mr Ward, the inventor of closed cases and the famous cultivator and lover of botany had, in his garden at Clapham, an artificial water course specially designed to grow the royal fern and other bog-loving plants within the narrow limits of a London garden. The Wardian case was a great boon to Victorian fern collectors, and to many other botanists since. Essentially it consists of a box with some glass incorporated but which can be tightly sealed. The plant is placed within and embedded into a layer of soil. During the heat of the day the water can evaporate both from the soil and from the plant itself (a process called transpiration). During the night, as the temperature falls, the water condenses on the inner walls of the case and falls as 'rain'. The specimens being transported were therefore kept in conditions as close to their natural environment as possible.

The Filmy Ferns

Tunbridge Filmy Fern

The slender delicate filmy ferns are among the smallest British ferns and can be confused with mosses. The fronds of the Tunbridge filmy fern *Hymenophyllum tunbrigense* arise in the summer, but can persist over many years. Like the higher plants, there are both deciduous and evergreen ferns, and the pteridologist should not be deterred from going out into the field in winter. The fronds are pinnate and the pinnae, which form a wing on either side of the rachis, are described as pinnatifid. This means that although there are pinnae, they are not completely separated and have segments (or lobes) which are joined together at the base. These seg-

Tunbridge filmy fern.

ments are serrated and spinous.

This fern grows in matted tufts and can look more dead than alive. The veins on the fronds are its most obvious feature. The portions of frond surrounding the veins look like thin green films and account for the vernacular name. The generic name also accurately reflects this appearance since in Greek *hymen* means a veil and *phyllon* means a leaf. The specific name of *tunbrigense* is less accurate: it was first described from Tunbridge on the borders of Sussex and Kent, where it does occur, but in what must be regarded as an outlying station. Its distribution both worldwide in general and in Britain in particular must be defined as patchy, and Britain is its most northerly station. Despite the Tunbridge record, the distribution is biased to the west and the species can be locally common in south-west England, Wales, the western seaboard of Scotland and in the Western Islands. Favoured habitats include by the side of waterfalls, on tree trunks and in damp woods. It seems to demand shady areas and has an aversion to limestone.

The fronds vary in size from 2.5 to 7.5cm (1.0 to 3.0in), they grow almost erect and their overall outline is broadly lanceolate. Although green, they do have an olive-brown tint which can be diagnostic of the species. The fronds are very flat, a useful feature in a shade loving species. The sori tend to be clustered around the axis of a vein and towards the end of the pinnae. Each sori is protected within a pocket-like indusium which is initially green but, when ripe, assumes the brownish colour of the rest of the frond. The indusium consists of two round and somewhat compressed valves which have a notched and spinous upper margin. This two-valved indusium helps to separate the filmy ferns from the closely related bristle-ferns.

Wilson's Filmy Fern

Hymenophyllum tunbrigense demands shade and a certain amount of warmth and shelter. Its close relative, the Scottish or Wilson's filmy fern *Hymenophyllum wilsonii* has the same north-western biased distribution pattern but seems able to cope with more extreme conditions and is often encountered in bleak and exposed situations, thriving at altitudes above 305m (1,000ft). At one time this species

Wilson's filmy fern.

was known as the one-sided filmy fern and scientifically as *Hymenophyllum unilaterale*. This reflects the blades of the fronds of Wilson's filmy fern – they tend to be less divided and therefore have fewer segments, which are longer and more concentrated on the upper side, thus giving a distinctly lop-sided appearance. The brown coloured sori are also quite different from those of the Tunbridge filmy fern. They are pear-shaped and their apical margins totally lack the 'teeth' which are so typical of *Hymenophyllum tunbrigense*. Furthermore, the sori do not fit snugly along the plane of the blade, but jut out at an angle. There are also observable distinctions between the fronds, which in Wilson's filmy fern are a deeper green, are much more rigid, of a brownish tint and also tend to droop. This is caused by the arched rachis and the convex pinnae which all turn one way and thus add to the lop-sided appearance. Although the fronds are usually between 5 and 7.5cm (2 to 3in) long, they can become much longer than this due to the fact that they continue to grow over several years. In this event, the basal section can be brown, withered and dead looking, whilst at the growing point the new and delicate green fronds look very healthy indeed.

Killarney Fern

Like the Tunbridge filmy fern, the Killarney or bristle fern *Trichomanes speciosum*, although not common, is more widely distributed than its name implies. It has a strange distribution which is difficult to explain, occurring mainly in tropical areas of both Europe and North America. In Europe it tends to be confined to the western seaboards, especially areas washed by the North Atlantic Drift, an important branch of the Gulf Stream. Because of its rarity, the species was very vulnerable to ambitious fern collectors and many former sites were stripped of their stocks. Although a recovery has been recorded in some areas, particularly on Arran, *Trichomanes speciosum* still needs the protection it was given in the 'Conservation of Wild Creatures and Wild Plants Act' of 1975. Killarney in Southern Ireland has retained a healthy population, and this goes some way to justifying the vernacular name of a species which is easy to identify and is unlikely to be confused with other ferns. Even in Killarney, however, the plant hunters sought it out, as an extract from Anne Pratt's book *Ferns of Great Britain*, written in the 1880s, clearly shows.

'It is found in several stations in Ireland; the Turk waterfall near Killarney being one often visited by botanists, who have recorded the enthusiastic delight with which they have looked on the hundreds of delicate fronds which form green masses there. It was formerly seen by Mr Newman very near the waterfall, but the guide of the place has sold so many pieces of this rare treasure to visitors, that the plant is almost exterminated at that spot.'

The fronds resemble the filmy ferns in texture but are usually much larger and can reach lengths of 30cm (12in). They develop very slowly, however, often over a period of years. Fertile fronds do not produce sori until their third year, after which they wither away, but those which do not become fertile – and there are many – can last for many years. The species, however, is very dependent upon damp conditions and when these are available it can produce masses of what a Victorian writer described as 'a rich verdant drapery

Killarney fern.

to the wet rock'. The slightest exposure to drought causes the succulent frond to wither.

The fronds are so much divided that they are often described as three or four times pinnatifid. Both the pinnae and their pinnules are further divided into narrow segments, each with a central vein. These segments, unlike the filmy ferns, are never toothed. Some tips are round whilst others are notched, in which case the vein divides and one fork leads into each limb of the notch. The net effect is the development of an exceedingly beautiful fern, with a delicately transparent texture accentuating the intricate tracery of the fronds. Indeed, some observers have suggested that it is more like a seaweed than a fern.

The rhizome (or root-stock) is likewise distinctive, often reaching several metres in length. It winds and branches so as to form a mesh-like network over the rock and is covered with dark down-like structures. If examined under a microscope these do resemble hairs and perhaps may account for the name *Trichomanes* which means a bristle, although it is more probable that this name derives from a bristle-like receptacle which projects from the indusium protecting the sori. The old name for the species was *Trichomanes radicans*, the specific name certainly referring to the numerous roots which cover the rhizome and afford it extra anchorage. Old vernacular names were the European bristle fern and the rooting bristle fern, both of which are highly descriptive.

Crested Buckler Fern

This rare species, *Dryopteris crista*, is now unlikely to be encountered outside the Norfolk Broads where it can be refreshingly frequent in and around the boggy areas of the Broads. Drainage schemes involving acid run off from mines in other parts of England have sadly depleted what was already a restricted range recorded as long ago as the 1880s by Anne Pratt, who noted that:

'It occurs but in four counties of England, and is found at Bawsey Heath near Lynn; at Fritton, at Dersingham and Edgefield in Norfolk; on Woolston Moss, near Warrington, Lancashire; on Oxton Bogs, Nottinghamshire; at Wynbunbury Bog in Cheshire; and a few other similar localities.'

On a world basis it occurs throughout Europe even extending as far eastwards as Siberia and it also occurs on the eastern side of North America.

The creeping rhizome is a substantial structure often reaching 15cm (6in) in length and often appearing to be much stouter than its real diameter of around 0.6cm (0.24in) due to dead bases of old fronds and a tangled system of roots. During May the narrow erect fronds begin to develop. Each frond is about 60cm (24in) high and it narrows towards the upper part, although the apex is always round and never pointed, and the 30cm (12in) stalk or stipe has its base enfolded by pale brown blunt, egg-shaped scales. The pinnae making up the frond are narrow and triangular in outline, those at the base being broader although each has the same basic shape. All are clearly pinnatifid and each individual segment is attached along the length of its base and connected to the one behind it. As the frond matures, the pinnae increase in length and the pinnules become more widely separated. The central veins of the lobes are somewhat irregular and put out lateral branches. On the upper pinnae these lateral veins bear the circular cluster of sori, each of which is initially protected by a flat kidney-shaped indusium. Mature spores can be shed from August onwards but the fronds die back as a result of the first autumnal frosts.

Crested buckler fern.

The Marsh Fern

Few species can have had their names changed more often, or been more adversely affected by human activities than this fascinating species, *Thelypteris thelypteroides*. At various times it has been described as *Aspidium thelypteris*, *Lastrea thelypteris*, and *Hemestheum thelypteris*. The word 'thelypteris' has been a common factor throughout and was therefore a sensible choice for the generic name. Some fern books still in print give the name as *Thelypteris palustris*, indicating its preference for boggy areas, although modern taxonomists appear to have settled for the name *Thelypteris thelypteroides*.

This is a species which varies very little; although there is some difference in size depending upon the amount of water in the habitat, the fact that it is so restricted to

Marsh fern.

fenland renders confusion with other British species most unlikely. It is still widespread in England and Wales although declining due to the drainage of mires, but it has probably always been rare in Scotland and Ireland. It is spread throughout the temperate and tropical areas of the world, always recognised by its creeping rhizome and almost complete absence of any scales.

The rhizome (root-stock) is rather short, seldom more than 60cm (2ft) long, and gives rise to single fronds rather than tufts of fronds, which are erect, bipinnate and without glands. Barren fronds arise in May, the fertile fronds arising during July and August. The sterile frond seldom exceeds 60cm (24in) in height, of which half consists of a straw-coloured hairless stalk. The texture is soft to the touch and a delicate green in colour, the scales present in the first week or so soon disappearing. The main outline of the frond which can be up to 12.5cm (5in) wide is lanceolate and the widely spaced pinnae are set at right angles to the rachis, alternately along its length. Each is obviously pinnatifid and described as sessile which means lacking a stalk. The undivided portions called the segments are broad and flat, the lowest pair usually being substantially longer than the others.

The fertile fronds are considerably taller than the barren ones, stiffer and with a relatively stouter stalk. There are often considerably more pinnae on the fertile frond, they are set closer together and are much narrower, whilst the fact that they tend to become folded over the sori may afford these important structures a not insignificant degree of protection. The sori are grouped in two rows, one on either side of the midrib on the lower surface of the pinnae. At first some additional protection is afforded to each sorus by a rather thin kidney-shaped indusium which, however, soon withers away and is lost. As previously indicated, this fern delights in marshy lands and botanists trying to establish its presence in old haunts would do well to search among the heathers, sphagnum mosses and flowering plants such as sundew *Drosera rotundifolia,* bog asphodel *Narthecium ossifragum* and the louseworts *Pedicularis sylvatica* and *Pedicularis palustris* (the red rattle).

Pillwort

All the ferns described so far in this chapter may be regarded as marshland plants. Only this species, *Pilularia globulifera*, and the water fern *Azolla filiculoides*, a description of which follows, can be regarded as anything approaching aquatic plants. Also known as pepper grass or pepper moss, the pillwort is widespread throughout Britain from north-eastern Scotland, but nowhere is it common, and it is decreasing rapidly in some areas because of extensive drainage programmes.

This plant never occurs in deep water, but often thrives in the shallows and in damp areas left behind after floods or heavy rain. The long, gangling, but quite thin and thread-like stem is hollow and only occasionally branched. Every centimetre or so small tufts of delicate roots descend into the mud and immediately above these, from the upper surface of the rhizome, a number of thread-like fronds (usually between two and six) arise. These are about 5cm (2in) long and a bright, startlingly attractive green. When young they are coiled like a shepherd's crook, but they soon uncurl to produce an almost military straightness.

Between June and August the spore bearing organs arise on short stalks at the base of the leaves. They are called sporocarps and each is 0.05cm (0.02in) in diameter and shaped like a peppercorn. Their resemblance to a pill has also been observed in many countries. In France they call it *la pilulaire*, the Italians and Spaniards both know it as *pilularia.* In Holland it is *pillenkruid* and in Germany *pillenfarrn.*

The sporocarp is at first green, but as it matures the colour is almost black. Each is

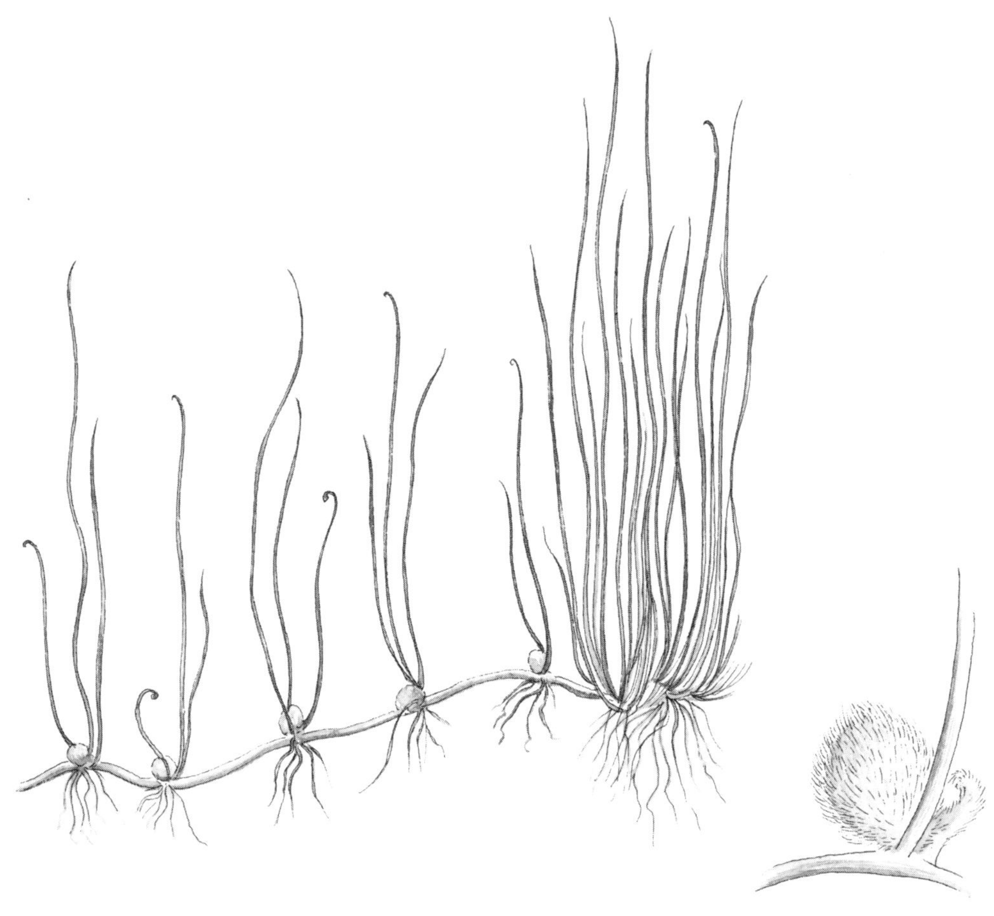

Pillwort.

divided internally into four compartments containing a sorus. Those in the upper compartments produce microsporangia, which in turn produce large numbers of microspores. The sori in the two lower segments give rise to megasporangia, each of which produces only one megaspore. The internal tissues of the sporocarp become mucilaginous when ripe and then absorb water. This causes them to burst and release the sporangia, a process which doubtless accounts for this species' total reliance on damp habitats. These it finds in western Europe from southern Scandinavia and the Soviet Union, and as far south as Portugal.

Water Fern

This, *Azolla filiculoides*, the only truly aquatic fern found in Britain, is a species almost certainly introduced from tropical America. Although it is a small plant it can mass together to produce dense floating masses in ponds and slow moving streams, such as those on the Somerset Levels. The stem of each individual plant is slender, much branched and seldom more than 2.5cm (1.0in) long. Long roots dangle from the lower surface whilst from the upper surface a number of overlapping fronds arise. Each of these is around 0.1cm (0.04in) long, often reddish in colour and bearing numerous one-celled hair-like structures. Each frond is divisible into two lobes, the upper one being green, but with touches of red, whilst the more delicate lower lobe is colourless. It is beneath this lower lobe that the sori develop, each covered by an indusium which is translucent. Like those of the pillwort, but unusually among the ferns, the sori are of two types which are termed microsori and megasori, depending upon whether they contain microsporangia or megasporangia.

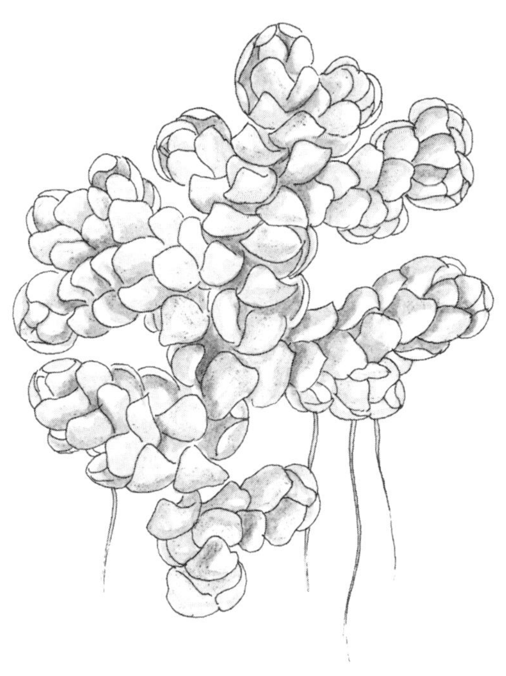

Water fern.

Another unusual feature of the water fern is probably related to the fact that it actually lives *in* rather than merely *around* water. Its floating fronds are perfectly capable of manufacturing carbohydrates from sunlight but there may well be a scarcity of mineral salts, especially nitrates, in the water. This would prevent the synthesis of essential proteins. Microscopical examination of the frond, however, reveals a cavity on the lower surface which leads into a cavity filled with slime. A blue-green alga, appropriately called *Anabaena azollae*, lives in the slime and can convert atmospheric nitrogen directly into nitrates, a process termed fixation. The algae have protection within the tissues of the fern which can utilise the nitrates when Anabaena cells die, and thus both partners may benefit. This association is termed a symbiosis – a living together for the mutual benefit of both.

One final point concerns the place of *Azolla filiculoides* in terms of the evolution of ferns. There are both reproductive and structural similarities with the *Hymenophyllacea*, the filmy ferns. It would seem a logical progression to evolve 'splash zone' species in the wake of aquatic ancestors, but whilst all scientists should be allowed to speculate they should all remember to proceed with caution.

6 Ferns of the Seaside

Many years ago when I first became fascinated by non-flowering plants I asked a local naturalist to take me on a fern hunt. I was astounded when he took me not to one of Lakeland's leafy woodland glades, nor yet on to the towering scree slopes, but to the seaside. There in the dune slacks he showed me such wonders as moonwort, adder's tongue and the lovely hay-scented buckler. There are inland sites for all these species but their existence by the sea underlines the point that ferns have conquered all environments. They are certainly worth hunting for with the salt breeze blowing through the air, the scent of ozone in your nostrils, the sound of oystercatchers piping on the sand and the waves crashing on the beach.

Moonwort

Moonwort, *Botrychium lunaria*, occurs in the temperate and cooler regions of both the northern and southern hemispheres. Few species can grow in so many countries, over an area which includes the whole of Europe save the Hungarian plain and along the coast of the Mediterranean. It grows throughout western and northern Asia as far as Japan. Australasia, including Tasmania and New Zealand, are populated by this fascinating little plant which also thrives in the arctic climate of Greenland and North America, including Newfoundland and Alaska. It spreads down through southern California, Minnesota, Vermont and into South America where it thrives in both Chile and Patagonia.

Its habitat is likewise very varied and it grows well on hills, grassy moors, heaths and stable sand dunes, often in the company of colourful flowers such as the harebell *Campanula rotundifolia*, rest-harrow *Ononis repens* and lady's bedstraw *Galium verum*. Moonwort is cer-

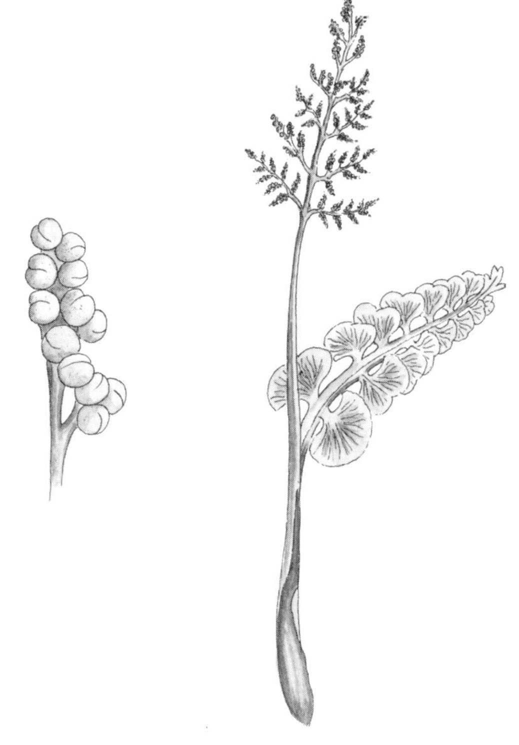

Moonwort.

tainly an uncommon species, but it is probably not as rare as it often appears because it is so small and often swamped by other vegetation. It also tends to grow singly, and never in colonies which would render it more conspicuous.

The frond of this little plant develops early in the spring, but is not particularly fern-like in appearance. Initially there seems to be a simple upright stem about 2.5cm (1.0in) high, which turns out to be a structure protecting two fronds, one fertile the other barren, which clasp each other. By June the moonwort has become fully developed but usually remains small, a few specimens reaching heights of 25cm (10in) but the majority less than half that. The colour is of a dullish yellow-green and around the base the traces of the scale-like sheath which once protected it can be seen. About half-way up it divides into two branches. One branch is pinnate and bears anything between three and eight pairs of fan-shaped – or perhaps crescent moon-shaped – pinnae.

The fertile spike is at least equal to, and more usually exceeds, the length of the sterile branch, often reaching 7.5cm (3in). It can be pinnate or bipinnate, each division bearing two rows of unstalked sporangia, one along each margin. Since all the pinnae point the same way, a fertile spike appears to be singularly one-sided. Initially the sporangia are yellow but this colour deepens to a dull brown when mature. By the end of July the sporangia will probably all have ruptured, the yellow spores within dispersed and the fronds withered away. The naturalist wishing to study moonwort must work hard during the period between April and June since there is little evidence of its existence at other times of the year.

The plant's pinnae account not only for its vernacular name of moonwort, but also for the incredible amount of folklore which surrounds it. We should not only express our wonder at the superstitions of our ancestors, but also realise how closely they must have examined the plants which surrounded them in order to note the shapes of the tiniest of their parts. Coles, a seventeenth-century antiquary, wrote that:

'It is said, yea, and believed by many, that Moonwort will open the locks wherewith dwelling-houses are made fast, if it be put into the keyhole; as also, that it will loosen the locks, fetters and shoes from the horses' feet that goe on the places where it groweth; and of this opinion was Master Culpepper, who, though he railed against superstition in others, yet had enough of it himselfe, as may appear by his story of the Earl of Essex his horses, which being drawne up in a body, many of them lost their shoes upon White Down in Devonshire, neer (*sic*) Tiverton, because Moonwort grows upon the heaths'.

This belief is also expressed in verse by Withers an early Stuart poet who, writing ten years after Gerard's death in 1622, pointed out that:

> *'There is an herb, some say, whose vertues such*
> *It in the pasture, only with a touch*
> *Unshoes the new-shod steed'*

Perhaps the words 'some say' shows some scepticism on the part of the poet, and disbelief certainly colours the account of moonwort written in 1687 by Turner in his book *British Physician*. Despite dismissing its value as 'smith, farrier and picklock', he remains convinced of its medical value and follows Gerard in stressing its value in the 'treatment of both green and fresh wounds'. John Gerard lived from 1545 to 1612, and Nicholas Culpepper, who lived from 1616 to 1654, did not always find it good sense to agree with his predecessor. On the moonwort question he sits well and truly on the fence.

'Moonwort is a plant which they say will open locks and unshoe such horses as tread upon it; these some laugh to scorn, and these no fools neither, but country folk that I know call it Unshoe the horse'.

Adder's Tongue

In the scientific name for adder's tongue, *Ophioglossum vulgatum*, the generic name Ophioglossum derives from two Greek words: *ophis* means a serpent and *glossa* means a tongue. The specific name vulgatum simply means common. These names are reflected in the names given to this fern in other parts of Europe. In France it was *langue de serpent*, in Italy *lingua serpintina* and the Swedes call it *laketunga*. In Germany it is called *natterzunglein,* whilst the Dutch name *adderstong* is almost identical to the English.

Early medicine and superstition went very much hand in hand and Dioscorides, the famous soldier-doctor in the days of the Roman Emperor Nero was a trendsetter in this respect. He, and many who followed him, held that every herb was under the dominion of the sun or of one of the planets, the identity of which could be discerned by their 'signatures' which were shown somewhere in the anatomy of the plant. This method of using herbs was in vogue for centuries and according to the medieval doctrine of signatures, a plant bearing a particular signature could be used to treat an organ of the body bearing the same signature. All the Christians did was to suggest that their God had provided the signature, as was indicated by Turner, the Dean of Wells, at the time of Elizabeth I, who produced the first British herbal.

'God has imprinted upon the Plants, Herbs, and Flowrs, as if it were in Hieroglyphicks, the very Signatures of their Vertues, as the learned Grollius and others will observe; as the nutmeg being cut resembleth the brain; the *Papaver erraticum*, or Red Poppy resembleth at Bottom the settling of the blood, in the Plurisie, and how excellent is that Flowre in diseases of the Plurisie and Surfeits has sufficiently been experienced'.

In the context of the doctrine of signatures, adder's tongue was far too good to be missed, and in 1657 Coles wrote a book called *Adam in Eden* in which he writes that: 'out of every leaf it sendeth forth a kind of Pestal like unto an adder's tongue; it cureth the biting of serpents'. When Phebe Lankester compiled *A Plain and Easy Account of the British Ferns* in 1854 the old writers had not been abandoned, and it was reported of adder's tongue that:

'even now large quantities of it are gathered in the villages of Kent, Sussex and Surrey, and prepared according to the old prescriptions. A preparation called Green Oil of Charity is made from it, and applied to wounds; and Gerard says: "The leaves of Adder's Tongue stamped in a stone mortar and boiled in oyle of olive, and then strained, will yield a most excellent green oyle, or rather a balsam, for greene wounds, comparable to oyle of St

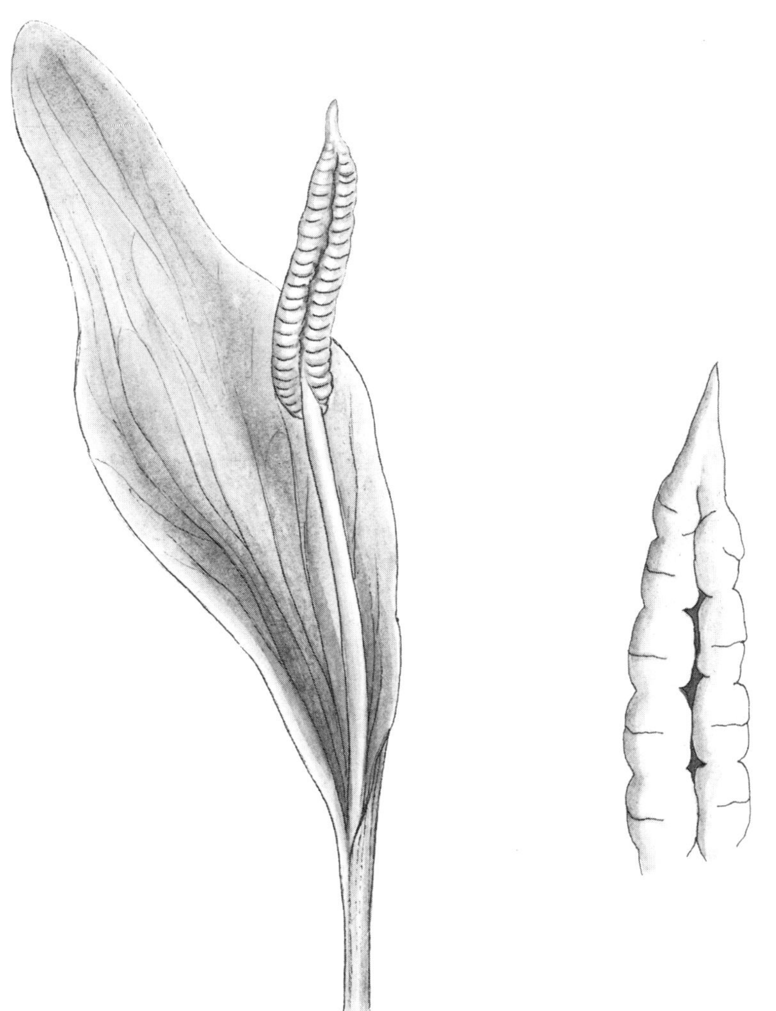

Adder's tongue.

John's Wort, if it does not far surpasse it". We incline to think that the only efficacious part of this ointment was the oil of which it was composed. Dr Lindley in his *Vegetable Kingdom* says: "The herbage of these plants is mucilagenous; whence the species have been employed in broths".'

Adder's tongue was on demand for the treatment of wounds – a so called vulnerary – right up to the beginning of the present century. Writers of doggerel weighed in with tongue thrust firmly into cheek. One such rhyme reads:

'For them that are with newts, or snakes, or
adders stung,
The seeking out an herb that's called
Adder's Tongue
As Nature it ordain'd, its own like hurt to
cure
And sportive did herself to niceties inure'

This firmly places the doctrine of signatures into the history books and finally removes it from the pharmacopoeia. But what about the biology of the adder's tongue fern? The yellow-brown, hairless rhizome is erect and fleshy and can be buried up to 15cm (6in) below the ground. The upper portion is clothed with the bases of old fronds and on the undersurface are numerous unbranched roots. Some of these elongate and behave rather like the stolons of the higher plants, producing apical buds from which new plants develop vegetatively. This may account for the often substantial colonies of adder's tongue which build up in suitable habitats. These include wet meadows, marshes, boggy stretches of woodland and, in the context of this chapter, the damp dune slacks between sand dunes. It can, however, be found growing in dry places. It is widespread in the Northern hemisphere but is rather less common in Scotland apart from areas of machair on the coast. It is worth mentioning at this point that botanists have suggested a subspecies which occurs on salt marshes among the short turf. This was called *Ophioglossum vulgatum ambiguum*, its separation based upon the presence of sterile fronds smaller than 3.5cm (1.4in) and having fewer than 14 sporangia on the fertile frond. More recently this plant has been raised to the level of a species and called *Ophioglossum azoricum*.

Unless a large colony of adder's tongue is cloned from the rhizome, the plant can be difficult to find because, like moonwort, it can be submerged beneath flowering plants such as water mint *Mentha aquatica*, great hairy willowherb *Epilobium hirsutum* and meadowsweet *Filipendula ulmaria*. It is, however, a case of once seen, never forgotten, mainly because it looks so unlike a fern. The single frond pushes through the ground in March or April, arising from the apex of the rhizome. Although it can reach 30cm (12in) it seldom looks so large as almost half of the structure is buried. This frond can look somewhat like lords and ladies, *Arum maculatum*, the broad sterile blade almost enclosing the taller and more slender fertile spike. The soft fleshy sterile blade is a succulent dark green and can be 30cm (12in) long and 5cm (2in) wide. Although there are veins visible there is no clearly defined midrib. The fertile blade is almost always longer than the sterile blade and at the apex of the fleshy stalk, called the spike, there is a substantial cluster of sporangia. When ripe, the yellow spores are released through transverse slits in the sporangia. By September almost all the adder's tongue plants have withered or are smothered by the higher plants. Once more, curious naturalists must be out and about in the spring and early summer if they wish to become acquainted with the ferns.

We cannot leave a consideration of this species without a brief reference to *Ophioglossum lusitanicum*, the dwarf adder's tongue. Basically a species confined to the South Atlantic and the Mediterranean, it has only penetrated as far as the Channel Islands and Isles of Scilly. Because of its habit of developing fronds among tufts of sea grasses on cliffs between November and February its num-

Top: mountain bladder fern; centre: Dickie's bladder fern; bottom: oak fern.

Top: royal fern; bottom left: crested buckler fern; bottom centre: water fern; bottom right: pillwort.

Left: great horsetail; right: marsh horsetail.

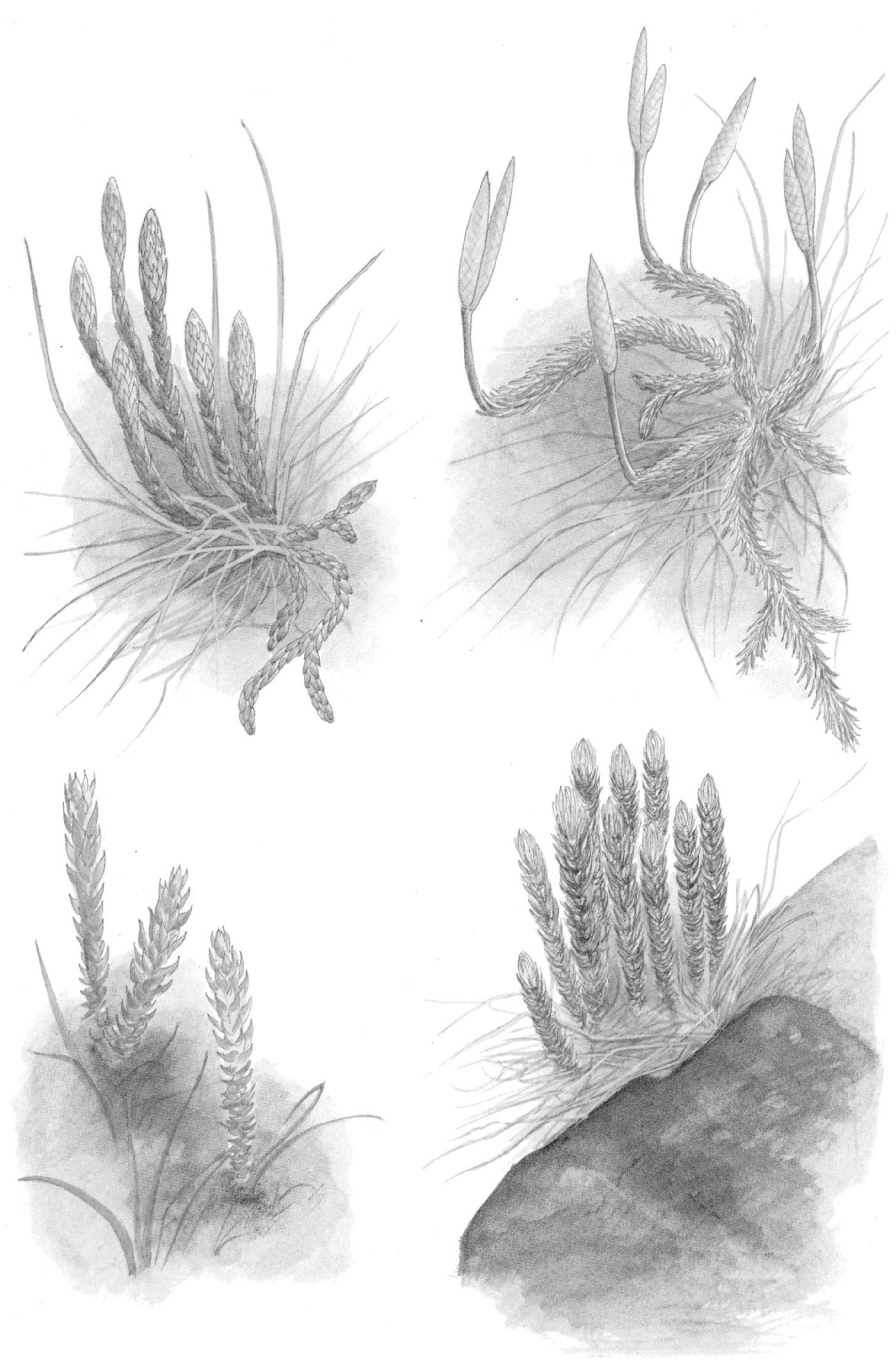

Top left: Alpine club-moss; top right: stag's horn club-moss; bottom left: lesser club-moss; bottom right: fir club-moss.

Top: maidenhair spleenwort; centre: hart's tongue fern; bottom: wall rue.

Top: scaly male fern; centre left: male fern; centre right: beech fern; bottom: lady fern.

Top: forked spleenwort; bottom: lanceolate spleenwort.

Top left: maidenhair; top right: sea spleenwort; bottom: hay-scented buckler fern.

bers have proved difficult to assess. Some feel that it may be present in Southern Ireland, but early records from Cornwall do not seem to have been substantiated and may have been sightings of very small specimens of *Ophioglossum vulgatum*. The small size of the species does nothing to help its discovery, the total length of the frond usually being less than 7.5cm (3in) and around half of this is stalk. The life history of the dwarf adder's tongue is very similar to *Ophioglossum vulgatum* but there are only six sporangia (sometimes fewer) arranged on either side of the spike which terminates in a thick blunt point. That of adder's tongue is much more pointed by comparison.

Jersey and Maidenhair Fern

Any naturalist with a passion for spectacular scenery and rare ferns will be amply rewarded if they go in search of these two rare species, *Anogramma leptophylla* and *Adiantum capillus veneris*, both Mediterranean and South Atlantic based plants and both belonging to the Adiantaceae family (*see* Chapter 4).

The Jersey fern is unique among British ferns in being an annual, although its distribution here is confined to the Channel Islands and is somewhat tenuous. It was discovered in the mid-nineteenth century and initially suffered from greedy collectors seeking either financial gain or academic prowess. Its growing season commences in the autumn, and fully mature fronds may be seen in the winter months on sheltered banks. During the hot summer the plant dies, leaving behind not a perennial rhizome, but tiny spores ready to germinate the following autumn. There are usually six fronds produced, each up to 15cm (6in) high. Those which develop first are rather small and drooping whilst the later growths are taller and more upright, but half the length of each is taken up by smooth brown stipe. The blades are a delicate pale yellow-green and initially are slightly hairy, but this characteristic has disappeared by the time they are mature.

Each plant is either bipinnate or even tripinnate, the pinnae being broad at the base, rounded at the tip and supported on a short stalk. The pinnules are themselves wedge shaped and divided into three lobes, each of which is also wedge shaped and notched at the tip. The short, linear, and quite small sori are lacking a protective indusium, but almost cover the whole of the undersurface of the frond with a deep brown mass of sporangia. The species needs to produce a vast amount of spores since these are its only means of survival – it has no perennial rhizome to rely upon should this reproductive phase prove unsuccessful.

The maidenhair fern *Adiantum capillus veneris* is another warm weather species which in Britain only occurs on sheltered western coasts washed by waters of the Gulf Stream. Although it has been recorded in the Isle of Man, on the west coast of Ireland and in Cumbria, there are no records for Scotland. It is more commonly encountered in south-western England and South Wales. Its natural preference is for sheltered limestone caves dripping with water, and close to the sea. It may have been more widely distributed in times past or perhaps carefully cultivated since the maidenhair fern was a popular plant with the herbalists.

Pliny, the Roman historian who perished at Pompeii in the eruption of

Maidenhair fern.

Vesuvius in AD 73, commented upon the smooth texture of the fronds and wrote: 'In vain you plunge the adiantum in water; it always remains dry'. The word itself derives from the Greek *adiatos* which means dry. Pliny was usually a reliable naturalist – within the context of his time – and doubtless knew many herbal remedies. Certainly the fronds were used in medieval Britain as a remedy for breathing problems, and when soaked in boiling water produce an aromatic odour, breaking down into a mucilagenous mass. In the Arundel manuscript it is written that: 'It mundifyeth (cleanses) the lunges and the breste, and casteth out of the body wykede materes in them'.

It is interesting that the North American Red Indians used maidenhair fern to cure their respiratory ailments. As usual the old herbalists made some exaggerated claims for the plant and asserted its value in the treatment of jaundice, swollen joints, pains in the chest and that: 'it stayeth the falling or shedding of the hair, and causeth it to grow thick, fair and well coloured'. They did recommend the addition of a herb called smallage to the hair restorer. This plant we know as wild celery *Apium graveolens*. It is interesting to see that the specific name *Capillus veneris* literally means hair of Venus.

Even the most junior of the apothecaries' apprentices would quickly have learned to identify the maidenhair fern since it possesses many unique features. It is an evergreen fern, especially when growing in sheltered positions. The bright green fronds arise from a much branched and creeping rhizome which is richly covered by scales. The very young fronds are delicate looking, red globe-like structures, which are then carried upwards on slender blackish-purple stems. The first branches arise half-way up the stalk which can be anything from 2.5 to 30cm (1 to 12in) long, depending on the habitat in which it is growing. The blade of the frond is ovate to triangular in outline, the upper portion being pinnate, the lower portion bipinnate, with the lowest pair of pinnae being the longest. All the pinnae are arranged alternately and supported on very thin stalks. Each segment (undivided portion) is fan shaped. In fertile fronds the margins of these segments are smooth, whilst those of the sterile fronds are obviously toothed. The characteristically round sori are positioned along the margins of the underside of the lobe and very close to the tip. There is no indusium to protect sporangia developing within the sori, but another protective device has been perfected. Part of the lobe bearing the sori is white in colour and is bent over to create a flap over the reproductive organs. The reproductive period is a long one, often lasting from May until well into September.

This is a species which has long been cultivated, and it grows larger in sheltered conditions, its delicate drooping foliage improving the appearance of many a floral tribute. It was, however, almost certainly first cultivated by the apothecaries to cure the coughs and sneezes of their clients.

Maritime Spleenworts

Lanceolate Spleenwort and Sea Spleenwort

The British spleenworts are typified by their straight linear sori, each protected by an indusium. All were used, as their name suggests, in the treatment of disorders of the spleen, as well as of the liver. The word derives from the Greek *asplenon* meaning 'of the liver'. The two species under discussion here, *Asplenium billotii* and *Asplenium maritinum* are both found mainly in coastal areas in Britain. *Asplenium billotii*, being a sub-Atlantic species, is totally confined to our warm western coasts, particularly in south-western England and Wales. There are records for Cumbria, the Western Isles, north-west Scotland and several from Ireland, but it is a very rare plant in these areas away from its main haunts. Elsewhere it occurs in the Mediterranean, North Africa and in warmer parts of Western Europe.

At various times the lanceolate spleenwort has been called the green lanceolate and also the spear-shaped spleenwort, whilst scientists first called it *Asplenium lanceolatum* and then *Asplenium obovatum* before settling upon its present appellation.

This, one of our most elegant ferns is seen at its best in Cornwall where, although the fern occurs inland, salt-laden breezes are evident even in the centre of the county. The rhizome is usually less than 2.5cm (1in) long, erect and densely covered by a mass of long, dark brown,

Lanceolate spleenwort.

pointed scales which look surprisingly like hairs. From the undersurface arise a number of dark warty roots, which penetrate deeply into the narrowest crevice in rock or wall.

New fronds arise during May, are usually producing spores by July but are evergreen, and withered fronds are usually still around as the new ones are developing. The bases of the fronds of several years give the rhizome a most untidy appearance long after the aerial portions have rotted away. Each frond seldom exceeds 20cm (8in) in the wild, but when cultivated or growing in particularly favourable conditions in the wild it can reach lengths of 45cm (18in) and look very spectacular indeed. Each is carried on a comparatively short, shiny brown stipe which may possess a few scales at the base, but is otherwise quite naked. The long blade of the frond is a rather bright colour of green, although never glossy, and varies in width from 2.5 to 10cm (1 to 4in), being described as lanceolate, although it widens in the middle before tapering to a point. The rachis is mostly green but it may sometimes be tinged with brown beneath. The frond is bipinnate with the large number of pinnae usually arranged in almost opposite pairs, which are horizontal apart from the last pair which are deflected towards the rhizome. Each of the pinnae is also broadly lanceolate in shape and the pinnules arising from them are egg shaped and have toothed margins, each tooth ending in a sharp point.

Close examination of the pinnules reveals a rather convoluted mid-vein giving off a series of branches leading towards each 'tooth' of the margin. It is close to these veins on the lower surface that the short linear sori are found, each protected by a snugly fitting white indusium. As maturity approaches, the indusium disappears and the sori run together, the spores being released towards the end of the summer.

The sea spleenwort *Asplenium maritinum* is a true maritime species with, however, a somewhat restricted distribution around the Atlantic coasts of Europe and North Africa. This is probably due to its demands for a rocky substrate on which to grow, the need to be struck by salt spray and also a degree of warmth. This accounts for its preference for the western coasts of Britain and its absence from much of the east coast, far removed from the effects of the North Atlantic Drift of warm water brought from the Gulf of Mexico by ocean currents.

The species, perhaps because of its mucilagenous appearance, has a long history of medical applications particularly in the treatment of burns. The rhizome (root-stock) is a short tufted structure densely clothed with dark brownish-purple scales. From it descend tough wiry roots which penetrate deeply into the clefts of rock or into the gaps in walls close to the sea. The fronds arise in tufts and can vary in length from 8 to 50cm (3.2 to 20in). The size varies considerably according to the situation in which it grows. The stalk, which can be up to half the length of the blade of the frond is red-brown in colour and, apart from a few scales at the base, is smooth and shiny. Two narrow green wings can be seen along its whole length. The blade (the green portion of the frond) varies from 7.5 to 30cm (3 to 12in) in length, can last for a year or more, and its thick green and leathery texture provides adequate protection from the salt-loaded winds driving in from the storm tossed sea. The frond is very variable in shape although it is de-

Sea spleenwort.

finitely pinnate, and the shape and arrangement of the individual pinnae are also the subject of disaccord. Some are ovate, others lanceolate and yet others being something between the two. In the lower areas the pinnae are arranged in opposite pairs, whilst those towards the apex are arranged alternately, but all are attached to the rachis by a short stalk.

The linear sori are grouped along the veins on the underside of each pinna and occupy the area between the crenated margin and the midrib. Initially they are protected by indusia but by late autumn these have withered and the bright red-brown masses of ripe sporangia are seen as bumps on the front. At this time the sea spleenwort is one of the most attractive ferns found in Britain.

Dickie's Bladder Fern

It was somewhat difficult to decide in which habitat to describe Dickie's bladder fern *Cystopteris dickieana* since it is so rare. The only description of the species is of a specimen found in a sea cave in Kincardine in Scotland. Botanists working in this field, however, do suggest that it may well be found at altitude in isolated spots in the Highlands of Scotland. Its generic name of Cystopteris refers to the red growths on the leaves which are the spores, and the most reliable way of distinguishing the species.

Hay-scented Buckler Fern

The hay-scented buckler fern, *Dryopteris aemula*, is basically a North Atlantic species, occurring on mountains of exotic places such as Madeira, the Azores and the Canary Islands. In Europe it occurs only in northern Spain, western France and the British Isles where its distribution is very much oceanic. Because there are so few inland records, the species is described in this chapter even though it occurs frequently in the damp woodlands on the west coast. Three important points of identification are the yellow-green fronds, the purplish stipe and the strong scent resembling hay from which it derives its vernacular name. This aroma is produced from tiny glands situated on the underside of the frond.

The hay-scented buckler fern has not been admitted to the British list of ferns without taxonomic argument. Hooker, in his monumental work *British Ferns*, expressed the view that it was merely a variety of the narrow buckler fern which was then named *Dryopteris spinulosa* and which has now been renamed as *Dryopteris carthusiana* (*see* Chapter 7). Hooker was also of the opinion that the broad buckler fern should also be regarded as a sort of superspecies along with the narrow buckler and hay-scented buckler.

It is certainly true that the hay-scented buckler fern does resemble smaller specimens of the broad buckler, but it can be distinguished in several particulars, including the scented fronds and the presence of brown scales which lack the dark central stripe typifying *Dryopteris austriaca*. The autumnal fronds of the broad buckler wither from the base upwards whereas those of the hay-scented buckler begin to wither from the tip.

The rhizome of the hay-scented buckler seldom exceeds 5cm (2in) in length and is usually erect, the older sections being clothed in a dense mat made up of the bases of old fronds. Wiry roots also descend from the older portions and provide a firm anchorage. From the younger section, which is covered with lance-shaped brown scales, numerous tufted fronds arise, each having the lovely scent of new-mown hay. The fronds vary in height from 15 to 60cm (6 to 24in) almost half of this being taken up by the slender, pale brown-purple tinted stalk which is clothed in scales. The blade varies a little in shape, being triangular to broad lanceolate. The upper surface is bright green above but rather paler on the underside, and both are liberally covered by tiny stalkless glands. It is a delightfully attractive species and its specific name of aemula meaning 'rivalling' presumably means

Hay-scented buckler fern.

that there are few other ferns to match it. Each blade can be from 7.5 to 22.5cm (3 to 9in) wide, tapering towards the pointed tip and is mainly bipinnate, although it can be tripinnate at the base. The pinnae are almost opposite the lowest pair, have an obvious stalk and are unequally triangular. Both these features alter gradually towards the tip until the upper pairs are lanceolate and sessile. The segments of each pinna are oblong with blunt tips and toothed margins which turn upwards, explaining why an old name for this species was *Dryopteris recurva*. The numerous and large sori are spread over the whole of the lower surface of the blade. Each is protected by a kidney-shaped indusium with several glands positioned around the margin. Fronds are evident from May, are mature during late summer, and wither when gripped by the icy hand of the first autumnal frost.

7 Ferns of the Woodlands

The ferns almost certainly evolved in warm humid environments and along with their allies created the habitat now referred to as woodlands. Until 5000 BC most of the temperate regions of the world were covered with trees, with many species of fern occupying the field layer of these botanically rich areas. No wonder, therefore, that woodlands are likely to contain more species of fern than other habitats. Those living outside the woodlands on the highland areas described in Chapter 8 are species which have had the trees that once protected them removed by human activities but have managed to hang on. Similarly, the ferns found growing on walls and described in Chapter 9 may be considered as the more adaptable of the woodland species. Among the species which are still basically woodland orientated are the beech, lady, male, scaly male, the narrow and broad bucklers and the hard fern. The chapter will conclude with a consideration of the ubiquitous bracken.

Beech Fern

Once known as *Phegopteris phegopteris* and *Thelypteris phegopteris*, the beech fern, *Phegopteris connectilis*, is widely distributed in the Northern hemisphere, but tends to avoid limestone areas. It is a very pretty and delicate plant which has also been called the mountain fern and the sun fern. It prefers to grow at altitude, but it thrives in the shady areas of woodlands, especially on steeply sloping hillsides. Its British distribution is obviously biased to the north and west, which by no means agrees with the distribution of the beech tree *Fagus sylvatica*. How the species became called the beech fern is not known with certainty, as there are no real physical similarities with the tree. There have been several attempts to have the vernacular name changed to one more fitting but most – fortunately – think it best to leave well alone.

The fronds which first appear in May vary in height from 15 to 60cm (6 to 24in) the stalk being around twice the length of the blade and slightly scaly at the base. The rhizome is up to 30cm (12in) long, dull brown in colour, liberally provided with black wiry roots, and coated with the leaves of old fronds. The older portions are covered with tiny hairs whilst the younger portions and any unexpanded fronds are protected by lance-shaped scales. The frond has a very distinctive shape, being triangular and tapering gracefully towards its point. The lower portion is obviously pinnated, each of the pinnae being narrow cut almost to the midrib, and also sharply pointed. They are usually arranged in pairs, but the lowest

Beech fern.

pair are positioned at some distance below the others and turn downwards towards the ground. This pair are joined to the stem by their midrib only, but the other lanceolate pinnae all point forwards and are joined to the stem along their whole length. They are also connected to each other in a pinnatifid manner (fused together at the base). The total surface of the blade, especially the midrib portion of the pinnae, is clothed in short silvery hairs. The circular sori, which have no protective indusia, are arranged in a row along either margin of the lower surface of each segment on the pinnae. When ripe, the sori give the beech fern a decidedly brown colour not unlike that of a decaying leaf of beech and this may be the reason for its vernacular name.

This is a very delicate species and soon perishes when exposed either to the hot summer sun or to the first frosts of autumn. *Phegopteris connectilis* is now included in the family Thelypteridaceae, but it was once the subject of some taxonomic debate and was classified with the polypodies. It was also for a time grouped with the oak fern *Gymocarpium dryopteris* and named *Gymocarpium phegopteris*. Until comparatively recently it was called *Thelypteris phegopteris*.

Lady Fern

'Where the copsewood is the greenest
Where the fountain glistens sheenest
Where the morning dew lies longest
There the lady fern grows strongest.'

Sir Walter Scott obviously knew this lovely fern, *Athyrium felix femina* very well and it still grows abundantly in moist woodlands, particularly those on acid soils throughout the British Isles. It also thrives in the temperate regions of both hemispheres, its lovely green fronds occasionally reaching 120cm (48in) in height, but subject to great, and sometimes confusing, variations. The scientific name for lady fern is felix femina and presumably alludes to its more delicate appearance when compared to the male fern *Dryopteris felix-mas*. Indeed at one time it was imagined that the two were the male and female forms of the same species. Although there are similarities there are also many differences in structure, particularly with regard to the sori, and the less scaly stalk and more divided root-stock of the lady fern. In the early days of a fern hunter's career, however, the lady fern is likely to be a difficult species, since even neighbouring fronds can show striking differences in shape and colour. However, for those who like to cultivate ferns and experiment in order to produce new varieties, this fern is indeed a popular lady and hundreds of different horticultural types have been produced. Nevertheless, the great majority of specimens will certainly fit the description which follows.

The stout and upright underground rhizome varies a great deal in size since it increases with age until it resembles a gnarled old tree trunk. The older sections are encased with the bases of old fronds, whilst the younger sections are covered in black lance-shaped scales. During the early spring the light green fronds arise in circular tufts from the apex of the rhizome. It is one of the loveliest sights, if you wander into a wood ringing with bird song and search among the wood anemones *Anemone nemorosa* and the sweet-smelling primroses *Primula vulgaris* for the unfolding scrolls of young lady ferns. Up to a dozen may occur in one clump, each looking like a miniature shepherd's crook. Each lanceolate frond has a stalk which can vary in colour from pale green to dark red and whose length is up to one third of the whole frond. This is bipinnate and up to 30cm (12in) wide, each pinna being lanceolate and gradually tapering towards the tip. The pinnae are arranged alternately, carried on a short stalk and are themselves divided into lobed pinnules which are sharply toothed and cut so deeply that they are described as pinnatifid (lobes fused together at the base).

The veining of the lady fern is one of its

Lady fern.

distinguishing features. A main vein meanders through each pinnule, but smaller veins branch off alternately and lead into the lobes, on the underside of which are two rows of sori arranged on either side of the midrib. The tiny sori, although they also show some variation, should be examined carefully, since they can provide a positive way to identify the species. Those situated near the tips of the pinnules are linear but those towards the base become kidney shaped, or would perhaps be best described as comma shaped. Each is protected by an indusium of the same shape which is only attached along one edge, the free edge being distinctly toothed. Those who wish to make a study of ferns must have an eyeglass which can be purchased very cheaply. Another useful acquisition is a binocular microscope, no more expensive than a reasonable pair of binoculars which no birdwatcher can do without.

Although the lady fern has not been of any interest to the apothecary apart from in association with the male fern, it was once used as protective packing during the transport of eggs and fruit, and in Ireland for packing fish.

Male Fern

Full details of the anatomy and physiology of this species, *Dryopteris felix-mas*, were described in Chapter 4, as were the problems associated with its complex taxonomy. In this chapter we shall therefore concentrate upon the extensive folklore connected with the male fern, which is widely distributed in the temperate regions of the Northern hemisphere and on some mountains in the tropics. In the middle of the nineteenth century, Phebe Lankester described the merits of the species thus:

'The medicinal properties of the male fern have been held in high repute for many ages; it is even now retained in our Pharmacopoeias as a vermifuge and was recommended by Theophrastus, Dioscorides and Galen. Lately it has been extensively employed as a remedy for tapeworm and with good effect. The attention of modern medical practitioners was probably first directed to it in consequence of its being the ostensible remedy of Madame Norisser of Switzerland, who sold her secret method of expelling tapeworm to Louis XVI for 16,000 francs. Tragus has a very curious passage on the subject of its curing wounds caused by reeds, and says that the antipathy of the male fern and the reed is so great, that the one will not grow in the company of the other. The same author also recommends a piece of the root of the male fern to be laid under the tongue of a horse that may have fallen sick from an unknown cause; by which means the disease will be expelled, and the horse restored to health. Even the homeopathic Pharmacopoeia recognises the use of the infusion of Male Fern roots as an agent in medicine.'

Schkuhr speaks of the ashes of this fern being used in bleaching linen and in the manufacture of glass, and also of an extract obtained from its roots used for tanning leather. Parkinson too mentions it as an ingredient in the making of a coarse green glass in France and in England, in his time. John Parkinson lived from 1567 to 1650 and was the Royal herbalist and author of a herbal called *Theatrum Botanicum* published in 1640. It is also worth reiterating that the use of the word 'root' means the root-stock or rhizome, the words being synonymous. It is certainly true that ferns were burned in potash pits, the resultant ash being the essential alkali in the manufacture of soap and glass. Ferns were increasingly used as the supplies of timber became depleted from the seventeenth century onwards when the industrial revolution gathered pace. Only when the inorganic chemistry industry developed did the practice cease and old potash pits can still be found in the remnants of the old forests of Britain.

Male fern.

Although its generic name has been altered several times from Polypodium, Aspidium and Dryopteris, it has always retained its name of felix-mas. Several continental countries also named it the male fern, obviously noting its more robust appearance when compared to the lady fern often found growing close by. In Italy it is *feli maschia*, in Spain *polypodio helecho masculino*. In France it is called *fougere*. Apart from sharing its uses, including those of cattle fodder, bedding and manure, the European countries shared the superstitions surrounding the male fern. Anne Pratt noted that:

'The young scroll-like fronds were formerly called Lucky Hands or St John's Hands, and believed, in days of darkness, to protect the possessor from all the ills of magic, the evil eye or witchcraft. The old German name of the fern Johannis Wartzel, reminds us of the usages common not alone in continental countries, but also in our own land. Not only was the yellow St John's wort dedicated to St John the Baptist, and burnt on Midsummer Eve in the fires raised in honour of the saint, but the delicate fern was duly gathered then, and sold to the credulous, who wore it about their persons, and mingled it in the water drunk by their cows'.

During previous chapters the fact that some emerging fern fronds look like bishops' crosiers has been noted, as has the doctrine of signatures. Is it not therefore highly probable that superstitious folk would believe that God had provided them with a sign to use against the devil? It could have been the young fronds which were hung around their necks on Midsummer's Eve.

Scaly Male Fern

Until recently this species, *Dryopteris pseudomas*, was called *Dryopteris borreri*, named after the British botanist William Borrer and until the 1950s it was regarded as a variety of *Dryopteris felix-mas*, but it is now clear that there are both genetic and structural differences between the two. The fact that hybrids can occur between the two adds to the confusion. The presence of large numbers of bright golden scales, from which the species is named, on the stalk and midrib is a good point of identification for *Dryopteris pseudomas*. The blunt untoothed segments on the pinna, each of which has a dark spot at its base, are good points of distinction although the dark spots are only evident on living specimens. The sori are arranged similarly to those of the male fern but they are smaller. Their indusia are always brown, never grey, even when young, and their edges are not flush but turned under.

The scaly male fern has been found distributed throughout Europe and south-western Asia. In Britain it is not quite so widespread as *Dryopteris felix-mas* although as a mainly Atlantic species it is particularly common in the west, especially north-western Scotland. When it does occur in the south and east it tends to be confined to woodlands growing on a base of heavy soils, especially clay. The presence of these conditions, plus a high level of precipitation, accounts for its appearance in some densities in the weald. High rainfall may account for its appearance in some mountainous regions where it grows in the cracks of stone walls and on the ruins of old hill farms.

The Buckler Ferns

Just as with the male and scaly male ferns, the narrow buckler fern *Dryopteris carthusiana* and the broad buckler fern *Dryopteris austriaca* have not proved easy to distinguish, especially where the two occur together. The beginner has not been helped by recent changes of name: the narrow buckler was once called *Dryopteris lanceolatocristata* and then *Dryopteris spinulosa*, the latter being still used in books at present in print. The broad buckler has likewise had a recent name change from *Dryopteris dilatata*.

Narrow Buckler Fern
This species has a circumpolar distribution and is widely distributed in the North Temperate zone. It occurs in both northern and central Europe as far as the Mediterranean. Although recorded from the Caucasus, the narrow buckler has not managed to penetrate into the more northerly areas of Russia and Scandinavia. It has been successful however in northern Asia and occurs as far south as Manchuria, whilst in North America there are records northwards of Virginia all the way to Alaska and Labrador. In Britain it is widespread and despite tending to become less common in more northerly areas, it is common in many of the wet woods of north-western Scotland, also the main

Narrow buckler fern.

resort of the northern buckler fern *Dryopteris expansa* described in Chapter 8. Up to the present no hybrids between the two have been recorded, but the search is on and examination of cells will no doubt come up with something. The Aspidiaceae family is anything but simple! Help is at hand, however, if we look closely at the structure of the narrow buckler fern.

The rhizome is characteristically prostrate and slender, seldom reaching more than 10cm (4in) and very scaly, with the older portions being clothed in the dark brown and dying remains of old fronds. The younger sections are covered with more healthy looking, ovate pale brown scales. These are points which distinguish this species from the broad buckler fern. There is also a significant difference in the way the fronds are produced, there being very few, usually not more than four, which never group together to produce a basket-shaped tuft. Each frond varies in length from 30cm to 120cm (12 to 48in) with about half this length being taken up by a thin stalk, which is dark brown at the base but lighter towards the top. It is covered at the base by scales which are also occasionally seen on the stalk itself. Each scale is cordate (heart-shaped) and pointed, but a major characteristic of this species is that although the scales are pale brown, they never have a darker central portion.

The blade of the frond is pale yellowish-green in colour, slightly leathery in texture, usually from 5cm to 17.5cm (2 to 7in) wide and lanceolate in outline. Each is bipinnate with between 15 and 20 pinnae on either side of the rachis. The lowest pairs have longer stalks and are arranged opposite each other, or almost so (technically this arrangement is called sub-opposite). As they approach the apex, the pinnae become arranged alternately and lose their stalks. The shape of the pinnae also varies as they ascend the frond, but they are usually widely spaced. The lowest pinnules on the lowest pinnae have short stalks, but as with the pinnae themselves, those close to the top lack any form of stalk and are therefore described as sessile. They are ovate to lance shaped and the pale brown sori are arranged on the lower surface of pinnules situated towards the upper part of the frond. They are arranged in two rows on either side of the midrib of the pinnule. Each sorus is protected by a kidney-shaped indusium but although these are always present, they can be rather inconspicuous on occasions. The spores are usually ripe between July and September, the fronds dying down quickly following the first autumn frosts.

Broad Buckler Fern

The broad buckler, *Dryopteris austriaca*, is an attractive and widely distributed fern which thrives in moist woods, especially around streams and waterfalls, and where there is plenty of leaf litter. The fronds vary in size depending upon where they are growing and good conditions may result in specimens exceeding 120cm (48in) in height. Specimens growing at almost 1,219m (4,000ft) on the mountain sides of Scotland may be less than one quarter of this size. Its distribution is circumpolar and it is found in Greenland at one extreme, and in South Africa at the other. In Britain it occurs almost everywhere except in a few areas of eastern England and central Ireland.

The stout upright rhizome is sometimes branched and covered with the dead, almost black remains of frond bases. The younger areas are clothed with lanceolate light brown scales. An examination of these scales will distinguish the broad buckler from closely related ferns. The scales of *Dryopteris austriaca* always have a very obvious dark brown central stripe. The scales of the narrow buckler fern are ovate in shape and never have the central stripe. The hay-scented buckler's scales also lack the line; they have uniformly brown lanceolate scales, although these structures may sometimes be described as laciniate which means that they are divided into narrow irregular segments.

In contrast to the narrow buckler, the fronds of this species which appear in early spring are numerous and are arranged in a basket shape. Almost one third of the length of the frond is due to the stout, scaly stipe, and each scale has the distinctive dark central stripe. The frond is broadly triangular, 15 to 40cm (6 to 16in) wide and bipinnate or even tripinnate. There are up to 25 pairs of pairs of stalked pinnae which are arranged alternately at the top but subopposite towards the base. The lowest pair of pinnae are unequally triangular in outline and are usually longer than the succeeding pair. The shape of the pinnae also alters gradually towards the top of the plant where they are lanceolate. The pinnules have short stalks, are ovate or lanceolate in outline, and are divisible into toothed segments, each tooth having a short spine-like point curved towards the apex. The numerous sori are arranged in two one on each side of the midrib of the segment, and are protected by obvious kidney-shaped indusia. These often have toothed margins and may be surrounded by small stalked glands. The spores are dark brown in colour and if examined under a microscope can be seen to be covered in tubercles. They are usually ripe by the beginning of August but some spores are still being shed by late September, although the fronds wither following the first frost.

Hard Fern

Although the distribution of this fern, *Blechnum spicant*, must be recorded as discontinuous, it occurs throughout Europe from the north of Iceland to Spain and stretches from Ireland in the west to the Caucasus in the east. Hard fern also occurs in North America from Alaska right down to California. It is found in most of the wet areas of Britain, although it does favour acid conditions. This causes it to be uncommon in limestone and agricultural areas. It is often a delightful member of the flora of a coniferous woodland.

Once known in the vernacular as the northern hard fern, ladder fern, rough spleenwort and herring-bone fern, this species has also had a number of scientific names which can cause the student some confusion. These names include *Blechnum boreale*, *Lomaria spicant*, *Osmonda spicant*, *Asplenium spicant*, *Acrostichum spicant* and *Osmunda borealis*. Its present appelation of *Blechnum spicant* has been in use for many years and looks to be secure. The word 'spicant' is a good description of the spiky nature of the fertile fronds.

Hard fern is an easily recognised species, having both barren and fertile fronds, as well as being evergreen. The barren fronds often lie prostrate on the ground, thus forming a rosette and bearing pinnae much closer together than the taller, more erect, fertile fronds. The fertile fronds are so finely divided into obvious segments that they bear an uncanny resemblance to a comb. Their appearance also explains why a Hebridean of my acquaintance

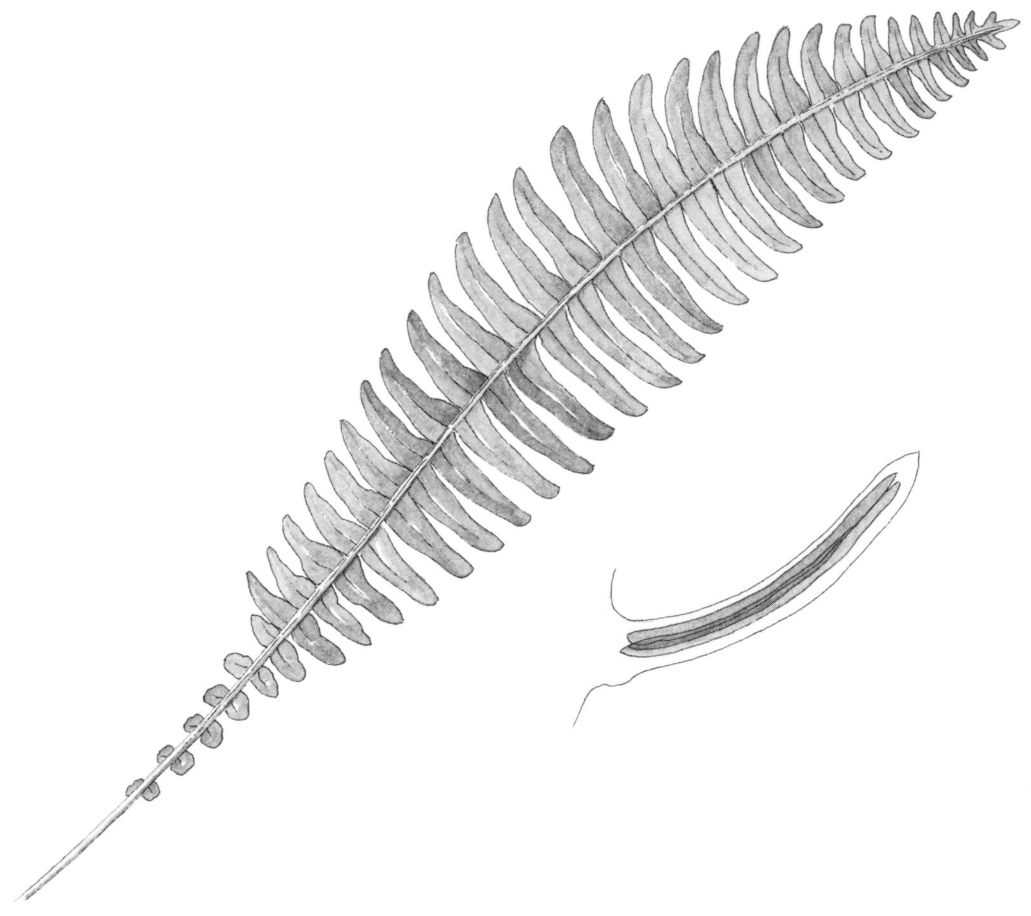

Hard fern.

called them 'herring-bones picked clean by the wind'. The fertile fronds are obviously pinnate. On the stalks of the fertile fronds which are approximately half their length are short scales, whilst those on the stalk of the sterile frond are long pointed scales. Both types are dark green in colour and seem to repel water from their shiny surface. The veining on the two fronds is similar except that on the fertile frond a long vein runs down each side of the midrib, and the sori are positioned on the underside, between these structures and the margin of the pinna. The sori are linear in shape and quite easy to see since they are almost as long as the pinna itself. Each is covered and protected by a similarly shaped indusium, which is white in colour and opens inwards.

The fertile fronds appear in the summer and die down in the winter but the sterile fronds persist right through the winter, and often delight the hardy naturalist when their shiny greenery is etched by a delicate coating of frost or snow. The rhizome is usually erect, about 7.5cm (3in) high but only about 0.62cm (0.25in) thick. It does, however, frequently appear to be much stouter than this, due to the remains of the bases of old fronds. The more mature areas produce masses of wiry roots, each liberally coated in brown hairs. The younger sections are covered by more delicate looking dark brown scales.

In view of its striking appearance it is perhaps something of a surprise to find that the hard fern has been rather ignored both by herbalists and by country superstitions. Rough spleenwort was, apparently, used by some herbalists but such quackery was condemned by John Gerard in his *Herball or Generall Historie of Plants* published in 1597.

'There be empiricks or blind practitioners of this age, who teach but with this hearbe not only the hardnesse and swelling of the spleene, but all infirmities of the liver also, may be effectually and in a very short time removed, insomuch that the sodden liver of a beast is restored to his former constitution againe, that is, made like into a raw liver, if it bee boyled againe with this hearbe. But this is to be reckoned among the old wives' fables and that also which Dioscorides telleth of, touching the gathering of spleenwort by night, and other most vaine things which are found scattered here and there in old books, from which most of the later writers do not abstaine, who many times fill up their pages with lies and frivolous stories, and by doing so do not a little deceive young students.'

It may well be that Dioscorides was referring to spleenworts in general, but Gerard's warning was still valid, even if he did not always heed it himself. We must admit, however, that it is of interest to know what our ancestors thought of the natural world.

Bracken

Few ferns have a more extensive folklore than the bracken, *Pteridium aquilinum*. This is the only British representative of the genus Pteridium, but it is certainly the best known of all our ferns. The name 'pteron' means a wing and alludes to the wing-like nature of the expanding fronds. Although it is found throughout Britain from sea level to heights of up to 609m (2,000ft) it is much happier in the warmer, wetter conditions found in the west and it does not grow particularly well on chalk. The countryman has known and made use of bracken for centuries and it has been given a variety of names including brake fern, eagle fern, adder spit, devil's hoof and even oak tree. Some used to say that when cut across, the pattern on the rhizome represented Charles II hiding in an oak tree. The bracken had, apparently, been so stamped to indicate a perennial reproach to it for failing to protect the monarch and forcing him to climb an oak tree. Yet another belief was that the initials 'J.C.' could be detected and this suggests that we are dealing with a holy fern, whilst to others the lettering was so indistinct that it was once used by country folk trying to make out the initials of their future spouse. It was perhaps most accurately referred to as the common fern and it is certainly the only species which forms huge closed communities, covering hillsides, heaths and open areas of woodland.

Bracken.

The underground rhizome is hard and tough, and lies so deeply in the soil that it is very difficult to dig up. The deeper area is much thicker than the upper region from which arises the frond. The lower region is obviously used as a food store. The best way to chop through the rhizome is with a sharp spade, when the two cut surfaces do indeed seem to resemble in shape an oak tree with outspread branches, or even the spread wings of an eagle. This accounts for yet another of its vernacular names and also for the specific name, since 'aquila' means an eagle. Bracken has provided the essential cover for game, and hunters have long waited for deer to emerge from its protection as indicated in the song 'The wild buck bells from ferny brake'.

In these days of easy living, the economical applications of harvesting bracken are not seriously considered, but in the 1850s it was still being used throughout

Britain as a manure. The glass and soap manufacturers in Scotland were burning bracken to produce the potash essential to their trade, whilst horses and cattle were bedded down on it. The stalks, which are very supple, were used for thatching, a practice which was recorded as early as 1349. In the Forest of Dean pigs were fed on the fronds, but these also proved an acceptable meal for hungry peasants. Phebe Lankester reported that:

'a botanical friend of our own, rather given to speculative devices, sent us one day a dish, consisting of the lower parts of the stem of this fern, cut off just below the ground, so as to retain the white delicate appearance of underground growth, assuring us it was quite equal to asparagus. It was accordingly cooked and served as sea kale or asparagus, and pronounced to be quite palatable, though not equal to either of the other named vegetables. It might, however, form a substitute for them, and, being so easily and inexpensively obtained, it is surprising that it does not oftener find its way to the poor man's table'.

Few insects seem to find bracken attractive as a food, but the caterpillar of the broom moth *Ceramica pisi* is an exception.

Bracken was also frequently kept in the medicine chest and was a cheap but very effective astringent, and was also used for dressing and preparing chamois leather. Another of its many uses was as a packing material, especially for fruit, but my great grandmother used it for transporting both crockery and crystal. In Saxon England the bracken was called 'fearn' and this has found its way into the names of many villages and towns, including Farnworth, Farnham and Farnborough, whilst some believe that the Farne Islands off the Northumberland coast were so named because of the extensive growths of bracken.

The wealth of folklore and proverbs in which ferns are mentioned almost certainly refer to bracken, and the following rhyme shows a good knowledge of nature throughout the seasons.

> *'When the fern is as high as a spoon*
> *You may sleep an hour at noon*
> *When the fern is as high as a ladle,*
> *You may sleep as long as you're able;*
> *When the fern begins to look red*
> *Then milk is good with brown bread.'*

Ferns seem to have been associated with invisibility and both Shakespeare and Ben Jonson refer to this belief. The bard writes of the 'fern seed by which we walk invisible' whilst Ben Jonson (1572–1637) is much more self-reliant and says:

> *'I hand no medicine, sir, to walk invisible*
> *No fern-seed in my pocket'.*

No one knows for sure why ferns became associated with invisibility but it seems safe to speculate that as all plants ought to bear flowers, their absence in ferns could indicate a magic power at work to conceal the blossoms. It was even suggested that ferns only flowered for a short period on Midsummer's Eve. In Staffordshire there was a belief that the bracken flower was very tiny and blue, but disappeared totally as the first flush of dawn pinked the sky. The Scots suggested that anyone who found a fern flower would not only be protected from evil but provided with treasure. Not all writers were gullible and in 1578 Henry Lyle wrote that ferns:

'beareth not flowers nore seed, except we shall take for seed the black spots on the backsides of the leaves the which do some gather thinking to work wonders, but to say the truth, it is nothing else but trumperi and superstition'.

The fronds of bracken begin to grow early in May and during the first stages they are reminiscent of a shepherd's crook which slowly unfolds as it pushes its way through the earth. They are quite delicate at this stage and easily killed during periods of late frost. The fully developed frond may vary from 60 to 210cm (2 to 7ft) and occasionally more, the overall shape being more or less triangular. The base of the frond tends to be clothed with yellowish hair-like structures which appear to contain a sugary substance found particularly attractive by ants. Picnickers soon learn to avoid stands of bracken fern!

The lanceolate pinnae are positioned almost opposite each other, those in the upper regions being flush with the rachis, whilst those lower down are carried on a very short stalk. The sori are arranged in a continuous line along the margins of the pinnules which branch from the pinnae. This means that bracken is usually described as bipinnate although the pinnules also occasionally have branches, in which case the frond is defined as tripinnate. Each sorus is heavily loaded with spores. A walk through an autumn stand of bracken, apart from being very heavy going due to the supple strength of the fronds, will leave the traveller covered in the rusty-coloured dust made up of millions of spores. There are, however, few more magnificent sights than a woodland glade ablaze with autumn bracken, or a hillside also clothed with bracken, upon which it is equally at home.

8 Ferns of the Uplands

Anyone with an ambition to see all our native species of fern in their natural settings will fall far short of the mark unless they are prepared to subject themselves to a hard climb. High on the mountain sides of Britain more than a dozen species are found and some of them occur nowhere else. Perhaps these areas were beyond the reach of the majority of Victorian fern hunters. The species to be described in this chapter include parsley fern, mountain or lemon-scented fern, the green, forked and alternate-leaved spleenworts, alpine lady fern, oak fern, limestone polypody, brittle and mountain bladder ferns, oblong and Alpine woodsias, the holly fern, the mountain fern and finally the very rare rigid and northern buckler ferns.

Parsley Fern

On many occasions during the researching of this book I have had cause to bless the writings of Anne Pratt, and only once has she made me sad. I had just returned from photographing parsley fern, *Cryptogramma crispa*, on the slopes of Coniston Old Man when I reread what she wrote about the species she called the curled rock-brake mountain parsley and rock parsley. She knew it also by the scientific name of *Allosorus crispus*. What made me sad was that I could then see why it was hard to find, apart from at high levels. Anne Pratt wrote:

'Many persons visiting the lakes at the north of England bring back with them a few fronds of this elegant little fern; and it is so beautiful in outline, and often renders the rocks so richly tinted by its green fronds, that it tempts even those who are not botanists to gather it. Southey, who describes it as the Stone Fir or Mountain Parsley, says it is the "most beautiful of all our wild plants, resembling the richest point-lace in its fine filaments and exquisite indentations".'

The parsley fern is a truly pioneering mountain species, thriving on screes and other often quite unstable areas in exposed situations. It may even help to stabilise the habitat and make it more suitable for other species to find a roothold. It occurs in most European mountains, even reaching the western areas of Siberia. It has also been discovered in Asia Minor and Afghanistan. It has an aversion to calcareous rocks and therefore thrives in the Welsh mountains, the English Lake District and the Scottish Highlands. Excessive collecting has reduced it to extinction in many former localities, although the occasional specimen is still encountered in Exmoor and at a few spots in

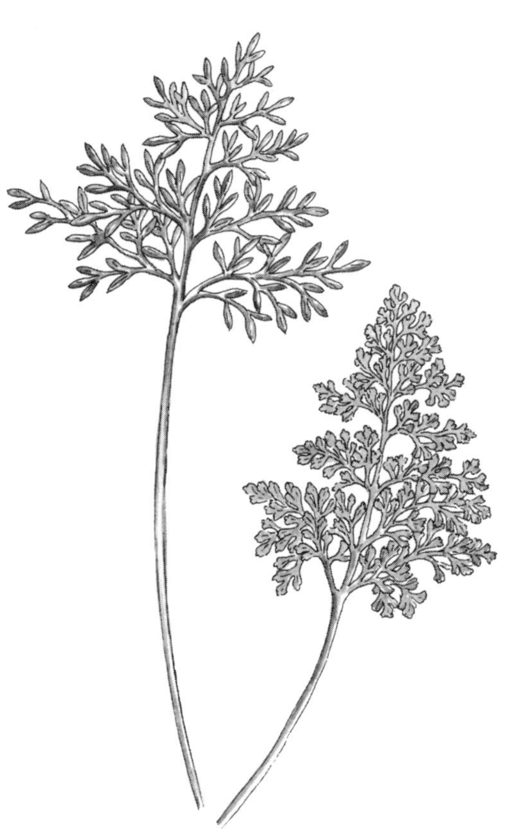

Parsley fern.

Ireland, including Ulster and Wicklow. Unlike many ferns this species does not show much variation and is easy to recognise.

The short, slender upright rhizome is often made to look quite bulky due to an accumulation of the dead bases of old fronds. From it arise large numbers of roots and two types of frond. The sterile fronds are grouped around the outside and may well protect the fertile fronds grouped within, which are much more erect and less numerous. The sterile fronds vary from 7.5cm to 15cm (3 to 6in) but around two-thirds of this length is taken up by the stalk which is smooth, pale green and surprisingly brittle. This characteristic is reflected in the specific name crispa. The blade of the frond is an ovate, delicate structure coloured bright green. It is very much divided, often described as quadripinnate, giving it some similarity to the leaf of parsley and thus accounting for its vernacular name. The blunt ovate pinnae are carried on a short stalk in almost opposite pairs. They are wedge-shaped at the base, lobed at the tip and have a tendency to overlap each other.

Although the fertile fronds are of similar size to the sterile structures, their stalks are considerably longer and can vary from as short as 10cm (4in) to as long as 30cm (12in). The texture of the fertile frond is best described as leathery rather than crisp, and it is of a much darker shade of green, although some specimens appear yellowish. It is invariably tripinnate, each segment being carried on a slender stalk, and almost linear in shape. The elliptical sori are found on the undersurface of the pinnule close to the vein endings. Initially the sori are quite distinct but expand as they mature to form a continuous strip running parallel with the margin of the pinnule. During the early stages, the frond may be curled and hide the sori, which may account for the generic name Cryptogramma, which indicates the 'hidden lines of sori'. When fully ripe the sori are clearly visible and the mass of brownish-yellow spores makes an obvious contrast against the green of the frond. It is no wonder that collectors found it attractive during the period between July and early September.

A third type of frond may occasionally be seen on the parsley fern and can be most confusing. In general its shape resembles a fertile frond but the flat segments are shaped like tiny oak leaves. All the three types of frond appear during May or early June and wither away following autumn frosts.

Mountain or Lemon-scented Fern

The name of mountain fern would suggest that there is no other chapter in which we could have placed this species, *Oreopteris limbosperma*. In recent years, however, the tendency has been to drop the word mountain and restrict its name to lemon-scented fern. This makes sense because the main requirement of the species is to have ground water flowing across its roots, particularly during the growing season. It gets this in upland areas where the rainfall is high but it can also be found along the sides of streams in lowland areas. It shows a distinct preference for acid soils, and can be very common in upland woods.

The lemon-scented fern's distribution is circumpolar and in Europe it occurs from Lapland down as far south as northern

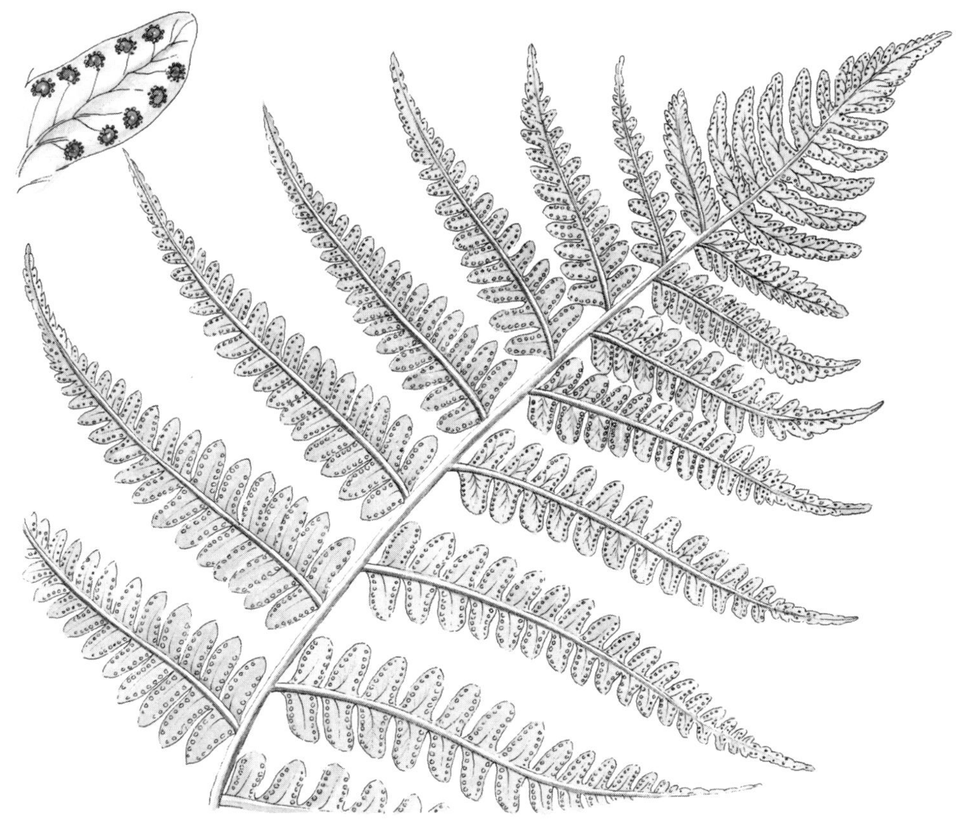

Mountain fern.

Spain. North-eastern Asia Minor, Japan, and North America from Alaska down to Washington, all have their colonies of the lemon-scented fern. This scent emanates from the crushed fronds, and it is certainly distinctive but whether 'lemon' is a correct description is open to discussion.

The stout and erect rhizome is around 10 to 15cm (4 to 6in) long, usually branched but sometimes growing in tufts, a condition described as caespitose. The older areas are clothed with the brown withering bases of fronds, but the younger parts are covered with brown scales which are usually ovate, although some are lanceolate. Dark brown unbranched roots descend from the under-surface of the rhizome, and from the apex arises a dense tuft of erect fronds which begin to appear late in the spring and die back in the autumn. The size of the fronds varies from 20cm (8in) to above 90cm (36in) but this does seem to depend on the suitability of the habitat rather than on any genetic variation within the species. The width can be anything up to 25cm (10in) but the overall shape of the frond is lanceolate and tapering at both ends. The colour also varies, but most specimens are bright yellow-green. The short stalk, never more than 20 per cent of the frond length, is covered with a few light brown scales.

The frond is pinnate, each pinna being lanceolate and arranged in opposite pairs. They are deeply pinnatifid and the pinnae towards the tip of the frond taper sharply to a point, whilst the lower ones are much more blunt. The whole lower surface of the frond is richly endowed with tiny golden glands and it is the release of the material within these which produces the characteristic smell when the frond is crushed. The sori are arranged in an easily recognised pattern, along the outer margin of the underside of each segment of those pinnae situated towards the top of the frond.

It is not easy to find the lemon-scented fern in many of the old textbooks of ferns since both the scientific and vernacular names have been changed so frequently. It has variously been known to scientists as *Thelypteris limbosperma*, *Thelypteris oreopteris*, *Dryopteris oreopteris*, *Lastrea oreopteris*, *Nephrodium oreopteris* and only in recent years as *Oreopteris limbosperma*. The vernacular names have been equally confusing and have included mountain buckler fern, scented fern, heath fern and hay scent. In view of the fact that there is a hay-scented buckler fern, *Dryopteris aemula*, already on the British list and described in Chapter 6, the continued use of the name hay scent for the species under discussion is, in some texts, confusing. Here is yet another indication that students should try to check the accuracy of scientific names used in texts, including the present one.

Green Spleenwort

As all plants are green, the epithet describing this attractive little fern, *Asplenium viride*, requires an explanation. In fact it refers to the green rachis which enables it to be distinguished from the maidenhair spleenwort *Asplenium trichomanes*, the rachis of which is dark brown. Many of the early botanists, including Sir Joseph Hooker, thought that the green-ribbed spleenwort was: 'perhaps an alpine subspecies of *A. trichomanes*, distinguished by its more flaccid habit, pale rachis, shorter, paler, and shortly stalked pinnae'.

Because there seem to be no folk names or uses for the species, we must assume

Green spleenwort.

that it has always been considered as a variety of the maidenhair spleenwort. The green spleenwort occurs in the Northern hemisphere and is circumpolar in its distribution. In North America it is found from Alaska, New Brunswick and Newfoundland and as far to the south as Wyoming and Oregon. In Europe it is distributed from the Arctic and Siberia down to the southern Alps, and in western Asia as far as the Himalayas. It is no wonder that the species makes light of growing in exposed positions on mountains and hillsides in Britain. Although it is not restricted to limestone, it does show a preference for this rock. The fern hunter active in non-limestone areas should keep a look out for the occasional outcrop and the green spleenwort will invariably have colonised the spot. I was once told by an old Dalesman who was an ardent bird-watcher that you do not find rare birds by chance. You first work out their favourite habitat and you then go there and wait. With ferns the advice holds good, except that you only have to search. The best places to discover the green spleenwort look very exposed until you see how the fronds are tucked into the rock to provide maximum shade and protection.

The slender rhizome is up to 10cm (4in) long but seldom more than 0.25cm (0.1in) in diameter, and is usually angled upwards which allows a tuft of fronds arising from the apex to emerge above the ground. The fronds vary in length from 5 to 20cm (2 to 8in) and appear in late spring. They have usually withered by midwinter. The naked stalk is never more than 50 per cent of the length of the frond, but attention should be paid to this part of the anatomy as it is crucial to the correct identification of the green spleenwort. The lower section of the stalk is brown but as it joins the rachis it remains soft and green along the rachis' whole length. In section both the stalk and midrib are round and there is never a winged shape as seen in the maidenhair spleenwort. Some botanists have made reference to the fact that the two species also show differences as they wither. In the green spleenwort the whole frond withers at once whilst in the maidenhair the pinnae are shed gradually, eventually leaving the midrib standing out like a stiff hair.

The blade is obviously pinnate and there are about thirty pairs of pinnae, each bright green in colour, glabrous, delicate in texture and ovate in shape. They are usually crenate towards the blunt tip and supported on a short stalk. The linear sori are situated on the undersurface of the pinna between the midrib and the margin and actually upon a secondary vein. The slightly toothed indusia fit snugly over the sori and are attached along one edge to the vein. The sori are ripe between August and October and are released when the indusia open inwards to reveal a mass of chestnut spores.

Forked Spleenwort

It is fascinating to discover how the interest in one scientific discipline can often provide unexpected information about another. A friend of mine was once showing me the spoil heaps of old lead mines in Yorkshire and remarked how often plants such as alpine penny-cress *Thlaspi alpestre* and spring sandwort *Minuartia verna* seemed to be attracted to these spots. Once my attention was drawn to this association, I began to take more interest and noticed that forked spleenwort, *Asplenium septentrionale*, also seemed to derive some nutritional benefit from these habitats. I congratulated myself until I read that way back in 1561 Gerard had made just the same observation!

The species is actually circumpolar in distribution, particularly on mountains, from Norway down to the Mediterranean and the Caucasus. It occurs in the western states of America and in northern Asia. In Britain there are records of it in the mountains of North Wales and Lakeland, which is understandable, but some isolated occurrences in lowland Scotland and Ireland which are less easily explained. There is no reason to doubt these records, however, because the forked spleenwort is not like any other British fern, and does

Forked spleenwort.

not vary in its shape.

The creeping branched rhizome gives off from its apex a cluster of fronds varying from 5 to 15cm (2 to 6in) in length which are so slender and forked that they resemble the antlers of a stag. These are thus so different from what one expects of a fern that they are often confused with the angiosperm species buck's-horn plantain, *Plantago coronopus*. The fronds do not, however, feel at all delicate and are quite tough and leathery. Pinnae do exist but they are very few in number – always less than five – one of which is terminal, with usually two more sited laterally. They can be lanceolate, but are more often wedged, tapering towards both the base and the apex which is somewhat acute. There is no distinct midrib and any number from one to five linear-shaped sori are arranged in parallel lines on each segment, taking up most of the lower surface. The indusia fit over the sori and open inwards. The brown spores are released any time between June and October but the fronds very often remain through the winter, and the brown shrivelled remains may still be present when the new green fronds appear the following spring.

The forked spleenwort is likely to remain a rare species in Britain, although there is evidence to suggest that it was at one time rather more common. Judging by what Anne Pratt informs us regarding events of the mid-nineteenth century this would seem hardly surprising. 'This plant grows in tufts, and, notwithstanding the diminutive size of the individual fronds, occasionally this forms large masses. Mr Newman says "At Llanrwyst, the tufts of this fern were very large; one of them was so heavy that after shaking out all the loose earth, I found it a very inconvenient load to carry even the single mile which I had to convey it. This tuft, consisting, I suppose, but of one rhizome, had upwards of three hundred perfectly vigorous fronds, besides at least an equal number of decaying ones, the relics of the previous year" '.

All one can say is that assuming that Mr Newman was a friend of such ferns then they certainly have no need of enemies!

We should certainly not leave the discussion of this species without reference to the very rare evergreen alternate leaved spleenwort. One even hesitates to give its scientific name because so much uncertainty surrounds its status. First it was accepted as a full species and named *Asplenium breynii*, which was subsequently changed to *Asplenium alternifolium*. Botanists have recently reduced it to the rank of a hybrid and named it *Asplenium x alternifolium*. They suggest that it occurs on lime-free rocks as a cross between the forked spleenwort and a rare mountain-based subspecies of the maidenhair spleenwort, which is named *Asplenium trichomanes subsp. trichomanes*. Obviously there is a great deal more work to be done on this rare fern which is only found in isolated places in the Welsh mountains and the Lake District. We are obviously no better conservationists than Mr Newman, since modern experts complain that the alternate leaved spleenwort has been 'overcollected for herbaria'.

Alpine Lady Fern

To find this fern, *Athyrium distentifolium*, is always a rewarding experience, coming at the end of a steep climb into the Scottish Highlands where it is tucked away among the snow-full gullies. To find the much less hardy lady fern *Athyrium felix femina* at these heights would not be possible, but the fertile fronds will bring instant distinction. Examine the sori which, in contrast to those of the lady fern, are not protected by indusia. Indusia in the alpine lady fern are either rudimentary or absent altogether. The spores of the alpine lady fern are also winged and are reticulately marked – they are not warty like those of the lady fern.

Not very long ago the alpine lady fern carried the scientific name *Athyrium alpestre* the specific epithet meaning 'of the mountains' – a good name for a species which appears to thrive at altitudes of up to 1,097m (3,600ft). Because of the absence of indusia there have also been a

Alpine lady fern.

number of attempts to classify it among the Polypodiaceae. Thankfully this has now been abandoned because this family is complicated enough as it is.

In some ways similar to the alpine lady fern is *Athyrium flexile*. This fern has no vernacular name and until very recently was described first as *Athyrium alpestre var. flexile* and then *Athyrium distentifolium var. flexile*. It is now recognised as a species occurring only in the mountainous corries of northern Scotland and it is characterised by reflexed fronds which seem to press down into the ground. The sori, instead of being spread evenly on the underside of the pinnae, are confined to the lower section. Both the sori and the fronds themselves are smaller than those of the alpine lady fern, but some botanists still consider that these may be due to variations in habitat. We may yet see *Athyrium flexile* reduced once more to a subspecies.

Oak Fern

Once called the pale mountain polypody and the tender three-branched polypody, *Polyposium dryopteris*, it is hard to imagine a set of more inaccurate titles for this lovely little fern, *Gymnocarpium dryopteris*. As we have seen in Chapter 4, it belongs not to the Polypodiaceae but to the Athyriaceae and dryopteris is hardly an accurate specific name either, as the oak fern has apparently no resemblance to the oak. We have already encountered the same confusion with the beech fern and the two occasionally are found growing together in upland woods. This is, however, the exception rather than the rule, as the beech fern requires liberal supplies of moisture, whereas the oak fern can exist in much drier conditions and is a true fern of sheltered spots on the mountainside. It does, however, show a marked aversion to limestone. The oak fern is not likely to be confused with any other fern except the limestone fern (*see* below).

In one sense the 'three-branched' part of the old name is accurate as the triple fronds are the most obvious characteristic of this plant which is attractively slender in form and sways about gracefully in the slightest of breezes. It is thin, smooth and fragile in texture and its colour is a really bright green, although much of the brightness is lost when exposed to the sun. The best specimens are found in the deepest shade. The fronds measure between 10 and 40cm (4 to 16in) and arise at irregular intervals from the slender creeping rhizome, which is black and glossy. The early development of the fronds was first described by Newman – the same man who ripped out the forked spleenwort from above Llanrwyst – as resembling three little balls on wires. This was an excellent description and has been used in almost every textbook since. These 'balls' unfold rather quickly during June and by the end of the month have produced the three graceful green branches, and also a mass of sori on the undersurface. These are carried on erect but surprisingly slender stalks which can be up to 80 per cent of the length of the frond. The stalk is usually green towards the top whilst the base is dark brown and has a sprinkling of brown oval scales, in contrast to the glabrous nature of the upper regions.

Oak fern is bipinnate, the pinnae being arranged in six opposite pairs, the lower pair being stalked, separated from those above, and by far the largest. It is these two pinnae which give the impression that there are actually three fronds on each main stalk. The pinnules of the lowest pair of pinnae are divided into lobes but are fused at their base. The pinnules on the rhizome side are longer than those near the apex of the blade. The sori are usually round, but some of the lowest of them may be rather elongated. There are no indusia and when ripe the sporangia tend to run together to form an almost continuous reddish-brown band. The spores ripen during July and August.

Limestone Fern

This species, *Gymnocarpium robertianum*, was once called the limestone polypody *Polypodium calcareum* but it is no more a polypody than the oak fern to which, as we have seen, it is closely related.

Oak fern.

The two were at one time considered to be varieties of the same species.

The differences between the two are as follows: the fronds of the oak fern are always glabrous whilst those of the limestone fern are densely covered by tiny glands, especially when young, and there are also brown lanceolate scales towards the base of the stem. In contrast to the slim black glossy rhizome of the oak fern, that of the limestone fern is much stouter and clothed in lanceolate brown scales and lighter brown structures which resemble strands of wood. The fronds of limestone fern are invariably larger, much more leathery in texture and a duller shade of

Limestone polypody.

green. In the oak fern only the lower pair of pinnae have obvious stalks whilst in the limestone fern the bottom three pairs are stalked. The specific name robertianum indicates another difference, as the crushed fronds of the limestone fern emit an unpleasant odour very similar to that of herb robert *Geranium robertianum*, a flowering plant of the crane's-bill family which grows on walls, especially in limestone districts. This brings us to the final point of distinction between the two species since, as its name implies, the limestone fern has a liking for this environment whilst the oak fern avoids these areas.

Gymnocarpium robertianum has a circumpolar distribution found as far north as Iceland and northern areas of Scandinavia and southwards as far as the Pyrenees, Corsica, Italy and the Balkans. In North America the range extends from Alaska and Labrador as far south as New Brunswick and Iowa.

In Britain the presence or absence of limestone appears to be the factor which affects distribution, rather than climate, and the limestone fern finds the grykes in limestone pavements ideal habitat. There are some records from eastern England which are accounted for by the number of disused railway buildings very rich in limestone-based mortar. In Britain at the time of writing I can find no records of hybrids between the oak and limestone ferns, but such crosses have been reported from Russia, Finland, Sweden and also, on several occasions, in North America.

Apart from the points of contrast listed above, the biology of the limestone fern is very similar to that of the oak fern and we can conclude this section by providing a list of names which have been used in the past to label the species. Vernacular titles included the rigid three-branched polypody and Smith's polypody. Other scientific names which have been used are *Lastrea robertiana* and *Gymnocarpium robertianum*.

Bladder Ferns

The bladder ferns belong to the same family as the limestone and oak ferns – the Athyriaceae – and two species in particular may be encountered in the highlands of Britain. These are the brittle bladder fern *Cystopteris fragilis* and the well-named mountain bladder fern *Cystopteris montana*.

Brittle Bladder Fern

The brittle bladder is the much more widespread species and provided that the ground is base-rich rock and that there is high rainfall it will thrive. Although this does mean it is ideally suited to mountainous habitats, it also means that it will grow well on walls and buildings facing the prevailing elements. On a world basis this fern may be regarded as a true cosmopolitan, having been recorded in North America from the Arctic, through Central and South America; Hawaii, New Zealand and Tasmania; Morocco, Ethiopia to South Africa; Greenland through Arctic Europe and Asia down to the mountains in tropical areas.

The bladder ferns derive their vernacular name from the fact that when immature the round sori are protected by indusia which are inflated and do indeed resemble a bladder. The generic name of Cystopteris comes from two Greek words also meaning bladder fern, and the specific name of fragilis is also accurate, referring to the brittle nature of the stalk. The beautiful little fern arises from a short-branching rhizome up to 5cm (2in) long and 0.5cm (0.2in) thick, its older sections clothed in a spiral of dead bases of old fronds. There are numerous tangled growths of roots, and the younger sections of the rhizome can be identified by a covering of slim pale brown scales. The lanceolate fronds which can be between 6 and 45cm (2.4 and 18in) long and up to 6cm (2.4in) wide arise as a tuft from the apex of the rhizome. The slender brittle stalk takes up between 30 per cent and 50 per cent of the total length. At the junction with the rhizome, the stalk is brown and scaly but the upper regions are yellowish and much smoother in texture. Each bright green frond is bipinnate, although occasionally tripinnate, and each pair of lance-shaped pinnae are arranged in opposite pairs and carried on short stalks, the last pair generally being widely separated from the others. The ovate segments

Brittle bladder fern.

of the pinnae bear two rows of sori, one on either side of the midrib, and are each protected by a bladder-like indusium which bends backwards during July and August to release the dark brown spores. These are comparatively large and coarsely spiny.

Mountain Bladder Fern

The mountain bladder fern *Cystopteris montana*, although once recorded in Wales, apparently in error, is confined in Britain to the mountains of Scotland. The suggestions of its presence in the English Lake District do not seem to have been substantiated. It has been recorded in the mountain ranges of both Europe and North America. There is no doubt that this is a rare species, but fortunately it is easy to recognise. It can be distinguished from the brittle bladder fern by its much longer creeping rhizome which does not sprout fronds from a tuft but at intervals along its length. The fronds are triangular in shape and seldom grow longer than 22.5cm (9in), with the stiff but slender stalk taking up at least half of this length. They are also always tripinnate. The only other species with which it can possibly be confused is the oak fern which also has a creeping rhizome and a triangular frond, but the oak fern's pinnae are much less dissected than in the mountain bladder fern and the oak fern also has naked sori. The mountain bladder fern develops its fronds during the late spring and they are withered before autumn gives way to winter, which is quite early on these very exposed mountain tops.

Woodsia

Both the oblong woodsia *Woodsia ilvensis* and the alpine woodsia *Woodsia Alpina* are very rare in Britain – so rare that both were overcollected and have had to be given legal protection by the Conservation of Wild Creatures and Wild Plants Act, 1975. The genus Woodsia can readily be recognised by its fringed indusia, but even more easily observed are its jointed stalks which do not occur in any other British genus. Woodsia commemorates the eminent British botanist Joseph Woods, who lived from 1776 to 1864.

Mountain bladder fern.

Oblong Woodsia

The fronds of oblong woodsia, once known as Ray's woodsia, arise from a tiny tufted rhizome, from the base of which grow black wiry roots. The upper surface of the rhizome is covered with the remains of old fronds, while new dull green fronds emerge in tufts. Each frond is lanceolate, pinnate and varies in length from 2.5 to 10cm (1 to 4in). The pale reddish-brown stalk is about half the length of the blade and jointed near the middle. The stalk and the rachis of the frond, into which it leads, are clothed with reddish awl-shaped scales and hair-like structures. There are usually between seven and fifteen pinnae on each side of the rachis, arranged almost in opposite pairs at the base but alternately towards the apex. Each pinna has a broad base and a rounded tip with deeply lobed and crenated margins. The two rows of sori are located on the undersurface of the segments. There is one row on each side of the midrib and close to the margin, each sorus protected by a cup-shaped indusium, each typified by long jointed hairs which enfold each sporangium within the sorus.

The oblong woodsia is a true arctic-Alpine species, found in Northern Europe from Iceland through Scandinavia, Switzerland and Northern Italy. In the context of its distribution, the specific name of ilvensis is difficult to understand, since the word is Latin for the island of Elba where the oblong woodsia does not occur. It does occur in the mountains of central Germany and also in the Carpathians, Crimea and Caucasus. The mountains of both Asia Minor and northern Asia also have many suitable niches for the species, as do Greenland and North America down as far to the south as North Carolina, Kentucky and Iowa.

Alpine Woodsia

The alpine woodsia *Woodsia alpina* is much more accurately named and was first described in 1785 by James Bolton (*see* bibliography). Although very similar to the oblong woodsia, it is certainly a discrete species and is identified by the absence of scales on both the stalk and the lower surface of the blade of the frond. With practice, differences in pinnae shape can also be detected, those of the alpine woodsia being almost triangular, and those of the mountain bladder fern being more lance shaped.

Alpine woodsia is an arctic fern which finds a similar hostile environment among the high mountain screes of the North Temperate zone in both Europe and North America. It occurs in Greenland and in North America from Labrador and Alaska southwards into the state of New York and western Ontario. The Ural mountains and northern Asia as far south as Altai also offer suitable niches for this tough little fern. In Europe it is found from Scandinavia and Northern Russia as far south as the Pyrenees, northern Italy and the Carpathians. Perhaps the oceanic climate in Britain may be rather too pleasant, thus explaining its extreme rarity and its confinement to inaccessible spots in North Wales, particularly in Snowdonia, and in Scotland where clumps have been found in Perthshire, Forfar, Argyll and on the Inner Hebridean mountains.

The rhizome of the alpine woodsia is a stout but much branched structure about 2.5cm (1.0in) long and 0.2cm (0.08in) in diameter. It is covered with leaf bases and tufts of roots are present in the undersurface. The tufts of fronds have pale reddish-brown stalks which, typically of the woodsias are jointed close to the middle and are up to 60 per cent of the length of the blades. The linear lanceolate and pinnated blade varies from 2.5 to 10cm (1 to 4 in) in height and from 0.75 to 2.3cm (0.3 to 0.9in) wide, and is a delicate pale green in colour. The sori are similar in shape and position to those of the oblong woodsia.

Holly Fern

Few ferns can stand up to high altitudes and exposure to bleak and blasting winds. The holly fern, *Polystichum lonchitis*, not only survives but thrives, its evergreen fronds peering out from the melting snow. The species is widely distributed in the alpine regions of Europe as well as in the arctic areas and the high mountains of Europe, Asia Minor, and the north of Asia as far as the Himalayas. Greenland also has its population and so do the highlands of North America down to Colorado in the west and Nova Scotia in the east. In Britain it is decidedly rare and confined to North Wales, the Lake District and the Scottish Highlands, seldom occurring below 305m (1,000ft). Some of the more southerly sites in Britain, usually on limestone, have been destroyed and others are threatened by quarrying.

In ideal conditions the fronds of holly fern may reach 60cm (24in) but they are often much more stunted than this. The stalk of the frond is comparatively short and covered, particularly towards the base, with ovate and lanceolate scales, always brown, but of variable shape. The dark glossy green blades are firm, rigid, leathery, erect and sufficiently prickly to make the reason for its vernacular name obvious. The young fronds appear early in the spring, pushing their way through the still green fronds of the previous year. They rise as a tuft from the apex of the short, thick, erect rhizome covered with the decayed bases of decayed fronds. Each frond is narrow and linear, tapering towards the apex and reaching its maximum width of around 5cm (2in) in the middle. The upper surface is quite smooth whilst the somewhat paler undersurface is liberally sprinkled with tiny pale brown and hairlike scales. The frond is pinnate with about 40 pinnae on either side, occasionally in opposite pairs at the base but usually alternately set. Each pinna is somewhat crescent-shaped and is twisted in such a way that they seem to overlap each other. They are usually irregularly serrated at the margins.

The large sori are invariably confined to the upper half of the frond and form two rows on the lower surface of the pinnae. One row is situated on either side of the midrib. The greyish and very conspicuous indusia fit over the sori and their margins are irregularly crenated or toothed. Ripe dark brown spores are released during the late summer.

The holly fern is so specialised in its habitats that it does not present too many taxonomic problems. Hybrids have, however, been noted in herbarium material and although they are hardly likely to come into the field observations of the naturalist, some brief mention ought to be made of them here. *Polystichum x illyricum* is the hybrid between the holly fern *Polystichum lonchitis* and the hard shield fern *Polystichum aculeatum*. *Polystichum x lonchitiforme* is the hybrid between the soft shield fern *Polystichum setiferum* and the holly fern *Polystichum lonchitis*. A hybrid has also been discovered between the hard and soft shield ferns and named *Polystichium x bicknellii*. (*see* Chapter 9)

Holly fern.

Mountain Male Fern

The Dryopteris genus of ferns is confusing because it is still very much in a state of flux. This is well demonstrated by both the history and the natural history of *Dryopteris oreades* the world distribution of which has still not been elucidated. In the past it has not always been allowed the status of a true species and has been called *Dryopteris felix-mas var. abbreviata*, then it was elevated to *Dryopteris abbreviata* and for a while it was removed from the genus altogether and called *Lastrea propingua*. It is now – for the present at least – given the name of *Dryopteris oreades*. For many years it was known as the dwarf male fern, which I find every bit as descriptive as the mountain male fern. It grows on screes, rock crevices and even on stone walls in all the highland areas of Britain.

Precise identification is not easy and

some careful measurements are required to separate it from *Dryopteris felix-mas*, but the following brief points may be of assistance. The rhizome, apart from being smaller, is also more branched in the dwarf male fern, the fronds are stiffer and almost always less than 50cm (20in), whilst those of *Dryopteris felix-mas* are much larger. There are also minute glands beneath the blade of the frond and pale brown scales on the short stalk of *Dryopteris oreades*. This scaly nature is carried on into the rachis of the frond. Although not easily seen in the field, the indusia covering the two to four pairs of sori (never more than seven pairs) are taxonomically very important. In *Dryopteris oreades* they are less than 1.0m (0.04in) in diameter compared to those of *Dryopteris felix-mas* which always exceed 1.5mm (0.06in) in diameter.

The Rigid Buckler and Northern Buckler Ferns

Both the rigid buckler *Dryopteris villarii subspecies submontana* and the northern buckler fern *Dryopteris expansa* are very rare species with particular demands regarding habitat. Only the most determined of fern hunters is likely to see either of them very often. Some books give the specific name of villarsii and although they are in a minority, they may be nearer the truth, as the fern commemorates the work of the French botanist Dominique Villars. He was born in 1745 and carried out his work in a period of real upheaval in the history of his country. He died in 1814.

The rigid buckler was at one time scientifically named *Lastrea rigida*. It is mainly a European species, being found in the Alps and in most montane habitats from southern Germany to the Pyrenees and even in parts of the Mediterranean and Bulgaria. There are records from Asia Minor and also Afghanistan. The British plants are restricted to areas of limestone pavement, and counts of the chromosomes in these populations have been found to be twice that of the alpine plants. This is why it is classified as a subspecies.

The British and continental plants have been artificially crossed and found to be fertile. The prostrate and slender rhizome can be up to 15cm (6in) long, 0.6cm (0.25in) thick and is often branched and densely clothed with dark brown bases of dead fronds. The younger areas are covered with bright brown lance-shaped scales.

The fronds arise during May as tufts from the apex of the rhizome and can be up to 60cm (24in) long, although they are frequently only half this length. The rhizome can be situated some distance below the cracks in the limestone pavement, with the fronds peeping out into the light. This situation may give protection from sheep who find the fragrant fronds which smell like balsam very much to their liking. The fronds are annual and die down as a result of the first frosts which in these exposed uplands can arrive before the end of September. Anne Pratt, describing the species under the name of the rigid fern, noted that the frond may:

'. . . assume one of two forms. In the one it is almost triangular, in the other lanceolate. It is twice pinnate, with narrow crowded pinnae, and pinnules which are blunt and oblong and cut again into broad rounded serrated lobes. The stalk is short, very full of scales . . . this has a pleasant fragrance, arising from the minute stalked glands which are scattered all over it . . .

Rigid buckler fern.

The mid-vein of the pinnules of the Rigid Fern is waved; branched veins issuing alternately from it, each becoming forked almost immediately on leaving the mid-vein. The lower branch divides again, each of the lesser branches running into a segment of the lobe. The upper branch – that is the branch nearest the top of the frond – bears the circular clusters of fructification about half-way between the mid-vein and the margin'.

By 'fructification' Anne Pratt means the sori, which are round and so crowded that the grey kidney-shaped indusia may actually overlap each other. The spores mature during July and August.

During the writing of this book I have frequently been surprised by the sheer energy of Victorian naturalists, including Anne Pratt, but especially Newman who despite his voracious collecting made observations which still hold good.

'I met with *Lastrea rigida* in great profusion along the whole of the great scar limestone district, at intervals between Arnside knot, where it is comparatively scarce, and Ingleborough, being most abundant on Hutton Roof crags and Farlton Knot, where it grows in the deep fissures of the natural platform, and occasionally high in the cleft of the rocks'.

Apart from the scientific names already given, the rigid buckler fern has also been named as *Aspidium rigidum* and *Lophodium rigidum*. One final point to mention is that it can be distinguished from the male fern *Dryopteris felix-mas* by its scent and by its lower pinnae, which are gradually diminished.

The alpine or northern buckler fern *Dryopteris expansa* was first recognised in 1895 as part of the well-studied flora on Ben Lawers in Perthshire, and named *Lastrea dilatata var. alpina*. It is regarded as an arctic-alpine species but similar looking specimens have been found in oak woodlands on the west coast of Scotland and in Wales. Its distribution is still being mapped.

There is a great deal of work to be carried out because of its very close relationship with both the broad buckler fern *Dryopteris austriaea* and the narrow buckler fern *Dryopteris carthusiana*. If you are interested in rarities, and how and when new species are recognised, you should join one of the societies whose addresses are included at the back of this book.

9 Ferns of Walls and Hedgerows

The study of ferns is not restricted to those with the health and strength to storm up steep screes on to mountain tops or with the resources and time to spare in order to reach remote spots. Ferns grow on walls, out of drains, peep out of hedges, and cover bridges over rivers and canals with their succulent greenery. There are several species to be found within a half hour's walk of any city centre, even those as busy as London, Manchester, Birmingham or Glasgow. Among the species which have adapted to life in towns and hedgerows alongside busy roads include polypody, hart's tongue, black spleenwort, maidenhair spleenwort, wall rue and the rusty back. All these species would appear on everyone's list of urban ferns. The hard and soft shield ferns may, however, cause a few eyebrows to be raised, but they are, in my view, true hedgerow plants, although never ubiquitous.

Polypody

A mature student of mine – advanced sufficiently in years to tell the truth – remarked during a seminar on ferns that 'polypodies are a bit of a mess at the moment'. How right she was – and is – because the taxonomists are having a frustrating, although doubtless fascinating, time with the genus, *Polypodium vulgare agg*. The work of M.G. Shivas is important here (*see* Bibliography) but it is beyond the scope of this present book to go into the details which Shivas elucidates. However, what we can do is to explain the general biology of the common polypody.

The genus Polypodium is basically tropical and consists of some 50 species of fern with rather pretty fronds arising from a creeping rhizome which has tufts of roots hanging down at fairly regular intervals. This accounts for the name polypodium which literally means 'many feet' and the rhizome does indeed look very like an animal. Another feature of the genus is the epiphytic mode of life of many of its members. An epiphyte is defined as a plant growing on a tree from which it receives support, but never feeds upon it. Epiphytes are common in all tropical forests.

The fleshy rhizome of the common polypody can be as long as 60cm (24in) and around 0.62cm (0.25in) thick. The older sections, although scarred with the bases of old fronds, are fairly smooth, the younger areas being clothed with reddish and markedly pointed scales. The roots are also tinged by reddish hair-like structures which are very tough and make the epiphytic plant difficult to remove. Polypody needs all this strength, as high winds whistle through the branches of trees or between the cracks of an ancient ruin

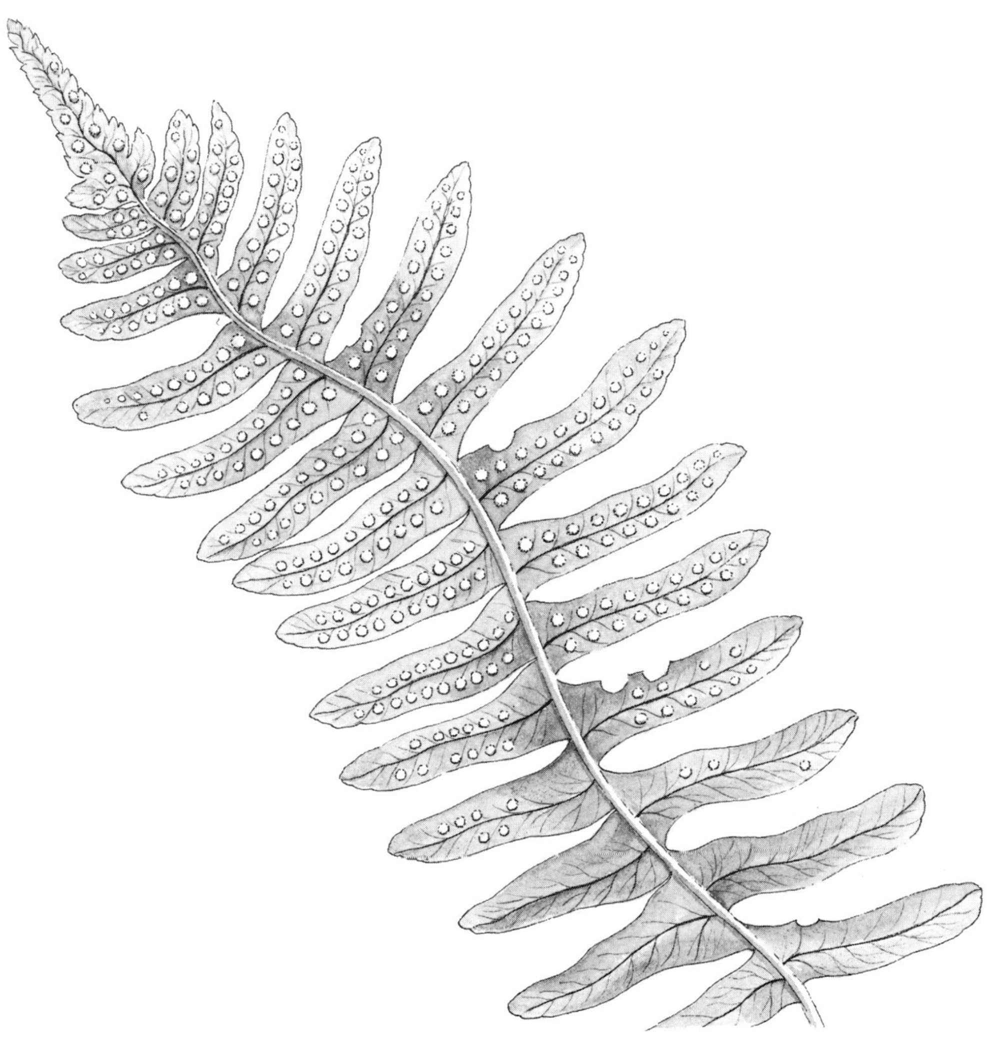

Common polypody.

colonised by the fern. The fronds arise in two rows along the rhizome, each being of an oval-oblong shape, measuring anything between 5cm and 75cm (2 and 30in) between 20 and 50 per cent of which is due to the smooth yellowish stalk. It is connected to the rhizome in a unique manner among British ferns. It is jointed by a special layer of cells so that, in the manner of the deciduous trees on which it often grows, it can be shed and leave a clean scar. This jointing is in no way similar to the jointed stalks which are unique to the woodsia ferns described in Chapter 8.

The evergreen fronds are smooth although leathery in texture and surprisingly tough, being little affected by either frost or drought. The fronds are deeply pinnatifid, the segments so produced having a broad base with either a blunt or acute apex and with the margins somewhat serrated. There is a tendency for the frond structure to vary and it is this tendency which has led to the taxonomic problems which surround it.

The numerous sori are large circular structures lacking an indusium and bright orange in colour making the polypody a delightful object during the period between June and early October when they are ripe. The sori are arranged in two regular rows on either side of the midrib, giving them a similar appearance to neat rows of buttons.

The species is widely spread throughout the zones of both hemispheres and grows throughout Britain in a variety of habitats including walls, hedgerows, buildings and trees. It has been given many vernacular names including adder's tongue in the New Forest, and golden polypody, golden lock and golden maidenhair in Kent. The last three obviously refer to the colour of the spores but the etymology of the name first listed escapes me.

The polypody has a long history of use in herbal medicine, being used as a purgative as well as for lung complaints and especially for treating whooping cough which was even more of a killer in previous centuries than it is now. The plant is therefore mentioned with respect in the old herbals. In the *Grete Herball* of 1526 the polypody is called the 'wall fern', whilst Turner calls it 'Brake of the Wall' in his *Libellus de re herbaria novus* published in 1538. In Lytes English version of Dodoens *Niewe Herball* of 1578 the word polypody is used perhaps for the first time. Gerard in his 1597 *Herball* calls it moss fern or everfern, by which he probably means 'evergreen'. Many old herbals also refer to it as the oak fern which has obviously led to confusion with *Gymocarpium dryopteris*.

In recent years *Polypodium australe* has been recognised as a polypody species preferring a limestone habitat and reaching its northerly limit on the island of Lismore above Oban. Because it is a Mediterranean and South Atlantic species it has been called the southern polypody. It is thought to have broader fronds which tend to be produced in the autumn, and the sori are linear rather than round. Other botanists recognise yet another species, produced as a result of hybridisation between the common polypody and the southern polypody and which has been named the intermediate polypody *Polypodium interjectum*. The search goes on for other species and subspecies. Polypodies may be a bit of a mess at the moment, but what a fascinatingly tangled web they weave!

Hart's Tongue Fern

Growing prolifically in the deep shadows of woodland glades, its bright greenery peeping from the clints and grykes of limestone pavements, or hanging grimly from the masonry of walls and bridges, the hart's tongue, *Asplenium scolopendrium*, is often overlooked because it does not look much like a fern – the frond looks much more like the leaf of a dock. It is particularly prolific in areas rich in lime and this explains why it grows so well in walls constructed using mortar and emerges from the edges of city drains. According to their habitat, the frond sizes can vary from 10 to 90cm (4 to 36in).

Hart's tongue.

The species is circumpolar in its distribution, being found throughout Europe as well as North Africa, the Azores and Madeira. It has been recorded through Asia Minor to Persia and also in Japan. In North America it occurs from New Brunswick down through the eastern areas as far south as Tennessee. In Britain it is evenly spread, apart from in central and lower Scotland, its absence from these areas being due to a combination of factors including low winter temperatures and lack of limestone.

The pale greenish-yellow fronds arise in bunches from a short stocky upright rhizome and each is divided by a very prominent central rachis. From this emerge delicate veins, which are initially at right angles to the midrib. Along these veins, on the underside of the blade, are the linear sori. These are grouped in pairs which are so close to each other that they often seem to be a single large structure. Each is protected by a linear membranous indusium attached along one edge to the vein of the leathery frond but with their free edges overlapping. When ripe, each sorus contains large numbers of reddish-brown sporangia and the millions of spores they produce are dispersed by the breezes of July, August and September. Once landed on a suitable habitat, the spores quickly germinate into tiny prothalli from which develop new fronds. The first fronds to develop are usually kidney shaped, but the subsequent fronds become more and more strap shaped.

In ancient Greece the species was known as *Scolopendrium* which means a centipede. When a fertile frond is examined closely, the sori do indeed resemble a centipede. Scientific names once applied to the hart's tongue – so called because its fronds are rather tongue-like – include *Phyllites scolopendrium, Asplenium scolopendrium, Scolopendrium crispum, Scolopendrium polyschides, Scolopendrium multifidum, Scolopendrium lacartum* and *Scolopendrium phyllitides*. The species was well known to the herbalists and was used in the treatment of burns and scalds, the strap-like fronds being laid across the damaged area. It was recommended as a dressing for ulcers and wounds whilst its renown as an astringent seems to have some substance in fact. It was also included in a compound produced by unscrupulous apothecaries and called the five 'capillar herbs'. The ferns used were the golden or common polypody, the common maidenhair, the common spleenwort, the wall rue and the hart's tongue. In view of the fact that so many varieties of the hart's tongue have been produced by horticulturalists, it is surprising that no subspecies have so far been described in the wild.

Black Spleenwort

This species, *Asplenium adiantum nigrum* is a true cosmopolitan and is widespread in Britain, particularly in the west where it is locally abundant in rock crevices, walls and stony banks, its evergreen fronds being easily visible throughout the year. Like all spleenworts it was used by herbalists in the treatment of liver and spleen conditions. Lyte, in his Niewe Herbal of 1578 refers to the species as the petty fern and also the black oak fern. Gerard in his Herball of 1597 scathingly noted that the black oak fern was used by some of his rival herbalists instead of adiantum of Lumbardie and, never one to be bashful, pointed out that these 'unlearned apothecaries . . . do err'. Apart from the names mentioned above, the species has also been called the black maidenhair spleenwort, but there are few vernacular names to confuse us. Likewise the scientific name of *Asplenium adiantum nigrum* remains as Linnaeus bestowed it in the mid-eighteenth century.

The creeping rhizome is about 10cm (4in) long and 0.5cm (0.20in) in diameter, the younger sections being covered with dark brownish awl-shaped scales which end in long tapered points. The organ is somewhat branched and tufted, the latter description also applying to the fronds which arise in June but which remain green for longer than a year. The fronds are between 15 and 30cm (6 to 12in) in height and are triangular or broadly lanceolate in shape, but are always widest at the base and the dark purple or black stalks are often quite as long as the leafy portion of the frond. It is these black stalks which have been responsible for the name of black spleenwort.

Each shining dark green frond which is paler on the undersurface, has a thick, firm texture with a prominent network of veins. The rachis of the frond is invariably green on the upperside, brown beneath and is usually winged along its length. It is difficult to decide how to describe the frond since it can be tripinnate or bipinnate at the base but becomes less divided towards the apex. There can be as many as 15 pinnae on either side of the midrib, usually arranged alternately and triangular in outline, the segments being ovate and toothed. The pinnae on the lower half of the blade are stalked, those further up being sessile. The linear sori are arranged in groups in the centre of each segment and attached to a vein, as are the similarly shaped protective indusia. The ripe sporangia produce an attractive rust coloured splash in the central section of the segment which can be apparent between July and October.

Common or Maidenhair Spleenwort

Anne Pratt refers to this species, *Asplenium trichomanes agg.*, as the common wall spleenwort which is a good name for this widespread fern. Two subspecies are now recognised in Britain, namely *Asplenium trichomanes trichomanes* which seems to be confined to Wales, the Lake District and Scotland, apart from the

Comparison of the two Subspecies of Maidenhair Spleenwort

Characteristic	*Asplenium trichomanes trichomanes*	*Asplenium trichomanes quadrivalens*
size of scales on rhizome	up to 3.5mm	up to 5mm
shape of scales on rhizome	lanceolate	linear/lanceolate
colour of central stripe of rhizome scales	red-brown	dark brown
colour of frond stalk	reddish-brown	dark brown
number of pinnae	14–28 pairs	16–30 pairs
texture of pinnae	delicate	robust (normally)
length of pinnae	2.5–7.5mm	4–12mm
usual shape of pinnae	suborbicular (almost circular)	oblong and parallel sided
normal number of chromosomes in the nucleus of the cell	72	144
habitat	non-calcareous rocks – upland	basic or neutral rocks – widespread
distribution	Europe only	Northern and Southern hemispheres – widespread

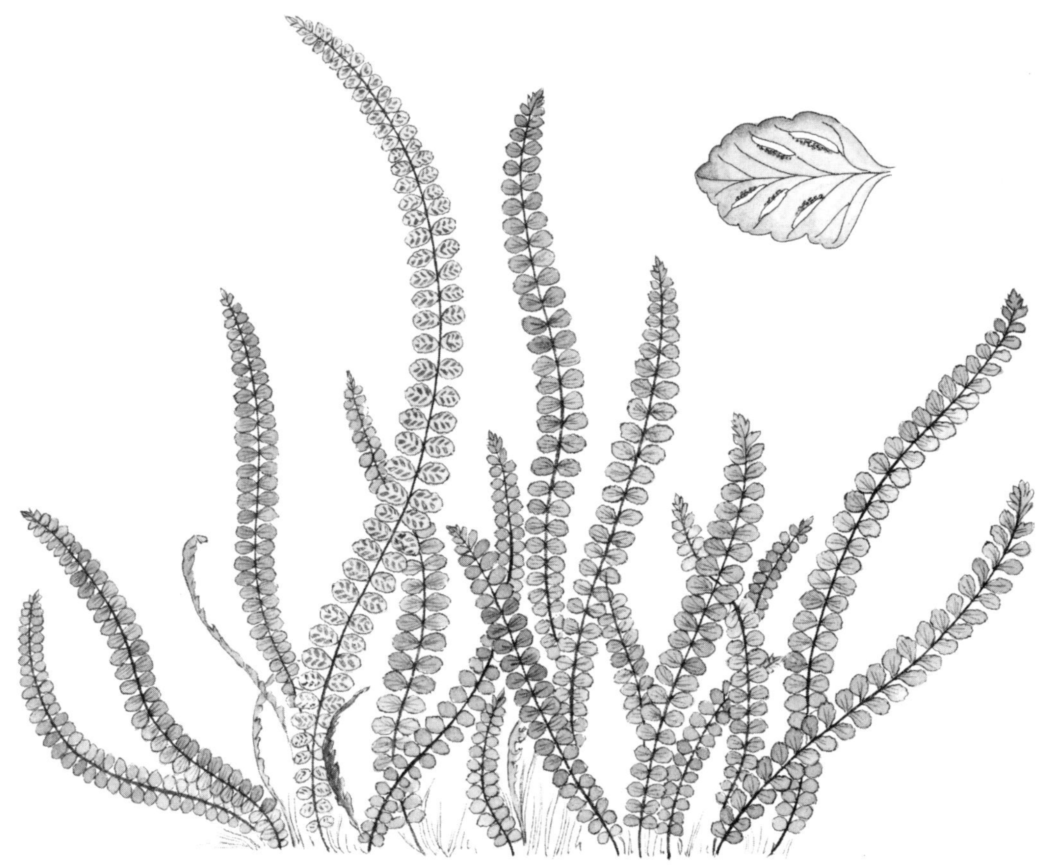

Maidenhair spleenwort.

occasional record from Northumberland, and the much more widespread *Asplenium trichomanes quadrivalens*. The latter prefers areas rich in lime, which are totally ignored by the former, and it is therefore frequently the one found growing on walls which are sealed with mortar. The main researcher in this area has been Dr J.D. Louis (*see* Bibliography). The two subspecies are not separable based upon one single characteristic, so the whole plant must be closely examined. The table on page 115 is compiled with reference to the work of Louis. Bearing the differences outlined there in mind, the description which follows applies to both subspecies.

The rhizome is creeping but rather stocky and seldom measures more than 5cm (2in), the younger areas being covered with brown scales. The fronds, each measuring between 5 and 20cm (2 to 8in) in length arise as a tuft from the apex of the rhizome. In the old days when this fern was called *Asplenium melancaulon* the fronds were brewed into a tea or mixed with honey to produce a syrup, both concoctions being recommended for the treatment of pulmonary infections. As the fern is evergreen, fronds would have been available the whole year round which must have added to its popularity. Even when the fronds do decay, the whole structure does not disappear. The pinnae fall and leave behind the black wiry looking midribs which may last for a year in this condition. They do look a little like hairs, but I do not know many maidens who would appreciate a *coiffure* of this nature. This characteristic is unique to this species and very useful to the field naturalist.

The narrow blade is linear, tapering slightly towards the base but more acutely towards the apex and never more than 2cm (0.8in) broad at its widest point. It is pinnate, and although the large number of pinnae can be arranged in opposite pairs, they may also on occasions be positioned alternately. The undersurface of each pinna carries a line of between 2 and 4 sori on either side of the midrib. Whilst young each sorus is covered by a delicate indusium but this falls off as the structures ripen and the sori themselves become swamped under a dark brown mass of spores. Sori may be seen from May onwards but it is usually late in the summer or even into autumn before the crop of spores has been finally liberated.

Wall Rue

This fern, *Asplenium ruta-muraria*, also called the rue-leaved spleenwort, is one species which seems to have found the habitats created by human activities more suitable to its well-being than its native haunts. The specific name of ruta-muraria suggests its love of walls, as does the similarity of its fronds (or rather pinnules) to the leaflets of the rue genus of flowering plants. For this reason the plant is not always recognised as a fern. In his *Names of Herbs* of 1548 Turner does call it wall rue but in his *Niewe Herball* of 1578 Lyte refers to it as stone-rue. Many old writers have referred to it as the churchyard fern and very few old walls are without a tufted cushion-like growth along the cracks and crevices.

In the 'wild' wall rue is generally distributed but it is more prolific in western districts. In areas around sea level the fronds tend to be quite small, but on the high hills and mountains of Derbyshire, North Wales, the Lake District and Scotland the fronds may reach lengths of 15cm (6in). This strongly suggests that its natural habitat is on mountains up to altitudes of 609m (2,000ft) where the short, scaleless root-stock pushes into cracks in the rock between cushions of mosses (and perhaps club-mosses) and lichens. The thin wiry roots seem to penetrate the rock itself and make it a most difficult plant to prize from its chosen niche – a point always made in the Victorian guides for fern-hunters. Wall rue grows particularly well on limestone and this is why it also thrives on walls constructed using mortar.

Distribution is described as circumpolar. It occurs throughout Europe and southwards to Algeria although this seems to be the only place in Africa where it does occur. Its range spreads eastwards into Northern Asia, and southwards into Afghanistan. In North America it is found from the far north down the eastern side as far south as Alabama.

New evergreen fronds begin to appear from late April, and are thick and leathery but a rather pleasing shade of dark green. The fronds persist through the following winter and the new greenery often pushes

Wall rue.

its way through the old fronds, which at that stage may still be green although they are more often brown and somewhat withered. The stipes make up some two-thirds of the length of the frond and are purplish-brown at the base but become green towards the apex. When young, they are clothed with hair-like scales. The fronds are bipinnate, the egg-shaped pinnules being arranged alternately and resembling small leaves, each supported on a short stalk with its upper margin irregularly toothed. The upper pinnae bear few, if any pinnules, whilst those lower down the frond may have as many as seven wedge-shaped pinnules. The veining on these structures does not branch from a definite midrib but forks in such a manner as to produce a fan-shaped segment.

The pinnules bear the sori which vary in number from two to five and are linear, each covered and protected by an indusium which varies greatly in shape. The sori may be visible throughout most of the summer, but ripe spores are normally produced towards the autumn. The ripe sori can produce so many brownish spores that they cover the indusia and give the impression that these structures are absent.

The apothecaries of old called this fern 'taintwort' (often mistakenly transposed to tentwort), which indicated its use in the treatment of rickets which was once called taint. Hybrids have been observed involving wall rue including *Asplenium x murbeckii* which results from a cross with *Asplenium septentrionale*, the forked spleenwort. *Asplenium x clermontiae* involves the subspecies of maidenhair spleenwort *Asplenium trichomanes quadrivalens* with the wall rue *Asplenium ruta-muraria*. It should be noted that both these hybrids are very rare indeed and it would take a dedicated expert to detect them.

Rustyback

The old name of *Ceterach officinarum* indicates that this unusual and exceedingly pretty fern, *Asplenium ceterach*, was useful in ancient medicine. We know this because the epithet *officinarum* meant that the apothecary recommended it, as his dispensary was called his office. We are again in debt to Anne Pratt, whose writings tell us that:

'The old Arabian writers said much in praise of its worth in complaints of the liver and spleen, and our herbalists eulogise its efficacy as an outward application to wounds. It appears to be the true spleenwort of the ancients, and the plant to which they attributed so great an effect in disorders of the spleen. The Cretan swine, when feeding upon it, were said to lose that organ altogether, and it was believed that, when taken to excess, the same injury was experienced by the human constitution. It has of late years been recommended as a good medicine in cases of jaundice. The fern is evergreen, and it grows to a much larger size in warmer regions than in our country. It seems, however, to be the same plant, owing its luxuriance to the climate. The author has seen a specimen of the scaly spleenwort brought from Madeira, in which some of the fronds of the tuft were fourteen inches long, though our native fronds are usually about three or four inches in length. The synonyms of this fern are *Grammitis cetarach*, *Scolopendrium cetarach* or *Natolepeum ceterach*'.

For the origin of the word spleenwort we must once more refer to the writings of another Victorian botanist, Phebe Lankester.

'The common name spleenwort takes its origin in a curious story – that in Cerito there is a river which divides two portions of land, the Cetarach growing abundantly on one side of the stream and not on the other. On the side where this fern grows, the pigs are said to have no spleen, but on the other side no such deficiency is recorded. Hence the name spleenwort or Asplenon.'

It should be noted that the letter 'A' before a word in latin signifies 'without'. Asexual for example means reproduction without sex, as in sporophyte generation. Among the many vernacular names given to this plant are brownback, scale fern, stone breaker, stone snap and stone fern, the last three indicating that it was once thought that the wiry looking roots of the fern were actually creating the cracks rather than merely growing into them.

The rustyback is a sub-Atlantic species confined to Europe and extending

Rustyback fern.

through the Mediterranean into central Europe and along the Atlantic coast to Britain, possibly reaching the limit of its range in Scotland. It has also been recorded from the Caucasus, North Africa and western Asia as far as the north-western Himalayas. There are no recent records from Madeira and Anne Pratt's specimen has been questioned by modern botanists. The more likely explanation, I feel, is that the observation was genuine, but the stocks were all removed by similar collectors.

The rustyback is common in Britain, particularly in the south-west and it thrives in the mild Atlantic Irish climate and where there is also a lot of limestone. It is easily recognised and cannot be confused with any other species. The rhizome is short and stout, a shape which enables it to become tightly wedged into crevices and walls. The old bases of dead fronds also add to the bulk and increase the efficiency of anchorage. The younger sections are clothed in pointed dark brown scales. The fronds arise as tufts and in Britain these vary in length from 2.5 to 25cm (1 to 10in), of which some 25 per cent consists of the dark stipe which is liberally clothed in scales. The blade of the frond is broadly lanceolate in shape and can vary in colour from quite a delicate green to a darker blue-green but the texture is invariably leathery. The frond is deeply pinnatifid.

The real distinguishing feature of this fern and from which it derives its common name is that the underside of the frond is covered by overlapping scales which even cover up the sori during the early part of their growth and make them very hard to locate. In young fronds the scales are silvery in colour but they become red with age, to produce a true rustyback fern. This scaly covering is a mechanism enabling the frond to cope with drought. During such periods, the scales can function like tiny muscles to turn the frond inwards and conserve water, only to open out again when moistened by the first drops of rain.

The sori borne in red-brown sporangia protrude between the scales when ripe which can be at any time between April and October. There is usually no indusium to protect the sorus, and even if one does occur it is always rudimentary.

Shield Ferns

The shield ferns are a genus constituted of over 200 species, but in Britain we have only 3 – the holly fern already described in Chapter 8, the hard shield fern *Polystichum aculeatum* and the soft shield fern *Polystichum setiferum*. The genus is typified by circular sori protected by a similarly shaped indusium which is unusual in having a central point of attachment, a condition termed 'peltate'. Another feature of the genus is the regularly arranged sori looking like lines of well-drilled troops. Indeed the word *polystichum* is Greek and means 'many rows'. Obviously the hard shield fern is harder to the touch than the soft shield fern, but other differences between them leave no doubt that they are discrete species. We should however be aware that the two will occasionally hybridise in the wild to produce *Polystichum x bicknellii.*

The hard shield fern arises from a short stout rhizome which does not differ markedly from that of the soft shield fern. The older sections are clothed with small pointed scales amongst which are a few rounder, larger scales, both groups of

Hard shield ferns.

Soft shield fern.

which tend to be rather darker than those of the soft shield. This is of course not a reliable characteristic when the rhizome is embedded in mud. When we come to compare the fronds, however, we are on much safer ground. The fronds arise in tufts from the apex but the hard shield produces fewer of them and they differ in both texture and colour. The stalk is very short, tends to be less than 16 per cent of the frond length and is clothed with similarly shaped scales to the root-stock. The blade of the frond is dark green or yellowish green and shiny above, but is much paler beneath and scattered with tiny hair-like scales. It narrows markedly towards the base and is a much narrower structure than that of *Polystichum setiferum*, although there is some overlap of sizes within the range of 5 to 25cm (2 to 10in) depending upon where the individual is growing.

The hard shield blade is pinnate or bipinnate, especially towards the base. There are around 50 short-stalked pinnae on either side arranged in opposite pairs in the lower areas but becoming alternate towards the apex. The pinnae are curved and their tips point towards the top of the frond. The upper pinnae are pinnatifid and sickle shaped and with toothed margins, which feel sharp and prickly. The small round sori are found mainly on the upper half of the blade, and are concentrated on the underside. Each is protected by a pale brown circular indusium typical of the genus.

The hard shield fern is less widespread than the soft shield and shows a distinct preference for limestone areas. It is widespread in Europe but does not penetrate into northern Scandinavia or into the north-eastern areas of the USSR. The soft shield fern is also a European species but is concentrated in western and southern regions penetrating as far as the Mediterranean. Both species are difficult to census in other areas of the world where there are many similar species. Here is yet another example of the work which remains to be done on ferns.

Typical of the soft shield fern are the soft, yielding fronds which are broader. They are much paler green and the pinnae look less crowded, probably due to the fact that the pinnules are smaller. The pinnules also have very short stalks and the bases are set at an obtuse angle. The large teeth have long, hair-like points. Its delicate fronds may account for its Cornish name of lace fern, although other vernacular names for the shield ferns are somewhat lacking.

Sir William Hooker, writing in 1861, was of the opinion that there was only one species of shield fern with several subspecies, the occurrence of which was dependent upon choice of habitat. More recent observations, particularly in Wales, have shown that the soft shield fern prefers a good depth of humus whilst the hard shield fern grows on more exposed rocky areas. We can therefore close this book by pointing out that naturalists must not only study the biology of ferns, but also go in search of British ferns in their habitats.

Useful Addresses

Botanical Society of the British Isles, c/o The Natural History Museum, South Kensington, London

Botanical Society of Edinburgh, c/o Royal Botanical Garden, Inverleith Row, Edinburgh

British Naturalists' Association, 23 Oak Hill Close, Woodford Green, Essex

British Pteridological Society, 46 Sedley Rise, Laughton, Essex

Council for Environmental Conservation, c/o Zoological Gardens, Regents Park, London

Countryside Commission, c/o John Dower House, Crescent Place, Cheltenham, Gloucestershire

Department of the Environment, 2 Marsham Street, London.

Field Studies Council, Preston Montford, Montford Bridge, Shrewsbury, Salop

Forestry Commission, 231 Corstorphine Road, Edinburgh

Institute of Terrestrial Ecology, 68 Hills Road, Cambridge

Nature Conservancy Council, 19–20 Belgrave Square, London

Royal Botanic Gardens, 47 Kew Green, Kew, Richmond, Surrey

Bibliography

Allen, D.E. *The Victorian Fern Craze*, Hutchinson, 1969

Alston, A.H.G. 'Notes on the supposed hybridisation in the genus Asplenium found in Britain', *Proceedings of the Linnean Society of London* 152 (1940) 132–44

Bellairs, N. *Hardy Ferns: How I collected and cultivated them*, London, 1865

Bolton, J. *Filices Britannicae*, London, 1785

Bower, F.O. *The Ferns (Felicales)*, 3 vols, Cambridge University Press, 1923–8

Clapham, A.R., Tutin, T.G. and Warburg, E.F. *Flora of the British Isles*, 2nd edn, Cambridge University Press, 1962

Copeland, E.B. *Genera Filicum*, Chronicle Botanica Company, 1947

Corley, H.B. '*Dryopteris felix-mas agg.* in Britain', *Proceedings of the Botanical Society of the British Isles* 7 (1967), 73–5

Crabb, J.A. and Shivas, M.G. 'Polypodium' in C.A. Stace (ed.), *Hybridisation and the Flora of the British Isles*, Academic Press, 1975, pp. 121–2

Druery, C.T. *British Ferns and their Varieties*, Routledge, 1912

Duckett, J.G. 'Spore Size in the genus Equisetum', *New Phytologist* 69 (1970), 333–46

'Coning Behaviour of the genus Equisetum in Britain', *Fern Gazette* 10 (1970), 107–12

Duckett, J.G. and Duckett, A.R. 'Reproductive Biology and Population Dynamics of Wild Gametophytes of Equisetum', *Journal of the Linnean Society (Botany)* 80 (1980), 1–40

Duckett, J.G. and Page, C.N. 'Equisetum' in C.A. Stace (ed.), *Hybridisation and the Flora of the British Isles*, Academic Press, 1975, pp. 99–103

Durand, H. *Field Book of Common Ferns*, 1928

Dyce, J.W. 'Growing ferns from spores', *Natural Science in Schools* 14 (1976), 43–5

Fawcett, E., ed. B.H. Lee *Lead Mining in Swaledale*, Faust Publishing Company, 1985

Foster, F.G. *The Gardener's Fern Book*, van Nostrand, 1964

Freethy, R. *From Agar to Zenry*, Crowood Press, 1985

Gerard, J. *Herball*, or *Generall Historie of Plants*, London 1597

Godwin, H. *The History of the British Flora*, Cambridge University Press, 1956

The History of British Flora. A factual basis for Phytogeography, 2nd edn, Cambridge University Press, 1975

Grounds, R. *Ferns*, Pelham, 1974

Heath, J. and Scott, P. *Biological Records Centre Instructions for Recorders*, Institute of Terrestrial Ecology, 1977

Hooker, W.J. *The British Ferns*, 5 vols., London 1861

Hyde, H.A., Wade, A.E. and Harrison, S.G. *Welsh Ferns, Club-mosses, Quillworts and Horsetails*, 6th edn, National Museum of Wales, 1978

Jahns, H.M. *Ferns, Mosses and Lichens of*

Britain and Northern and Central Europe, Collins, 1980

Jalas, J. and Sucominen, J. (Eds) *Atlas Florae Europeae (Distribution of Vascular plants in Europe) I Pteridophyta (Psilotaceae to Azollaceae)*, Helsinki: The Committee for Mapping the Flora of Europe, 1965

Jermy, A.C., Arnold, H.R., Farrell, L. and Perring, F.H. *Atlas of Ferns of the British Isles*, The Botanical Society of the British Isles and the British Pteridological Society, 1978

Jermy, A.C. and Harper, L. 'Spore morphology of the *Cystoperis fragilis* complex', *British Fern Gazette* 10 (1971), 211–213

Jermy, A.C. and Page, C.N. 'Additional field characters separating the subspecies of *Asplenium trichomanes* in Britain', *British Fern Gazette* 54 (1980), 112–113

Jermy, A.C. and Walker, S. 'Dryopteris' in C.A. Stace (ed.) *Hybridisation and the Flora of the British Isles*, Academic Press, 1975, pp. 113–118

Lankester, P. (ed.) *British Ferns*, Robert Hardwicke, 1854

Kay, R. *Hardy Ferns*, Faber and Faber, 1968

Lovis, J.D. 'The problem of *Asplenium trichomanes*' in J.E. Lousley (ed.), *Species Studies in the British Flora*, Botanical Society of the British Isles, 1955 pp. 99–103

Louis, J.D. 'Evolutionary patterns and processes in ferns' in R.D. Preston and H.W. Woodhouse (eds.) *Advances in Botanical Research*, vol. 4, 1977, pp. 229–415

Lowe, E.J. *Fern Growing*, Nimmo, 1845

McClintock, D. 'The ferns of the Channel Islands', *British Fern Gazette* 9 (1961), 34–7

Milne-Redhead, E. and Trist, P.J.O. 'A remarkable population of Ophioglossum vulgatum in Suffolk', *Watsonia* 20(4), (1975), pp. 415–416

Newman, E. *A History of British Ferns and Allied Plants*, 2nd edn., J. van Voorst, 1844

Page, C.N. 'The taxonomy and phytogeography of Bracken – a review', *Journal of the Linnean Society (Botany)* 73, (1976), 1–34

Page, C.N. 'The diversity of ferns: an ecological perspective' in A. Dyer (ed.), *The Experimental Biology of Ferns*, Academic Press, 1979, pp. 10–56

'Experimental aspects of Fern ecology' in A. Dyer (ed.), *The Experimental Biology of Ferns*, Academic Press, 1979, pp. 551–559

The Ferns of Britain and Ireland, Cambridge University Press, 1982

Payne, L.G. 'The Crested Buckler Fern', *London Naturalist* (1939), pp. 29–31

Perring, F.H. and Walters, S.M. *Atlas of British Flora*, 2nd Edn., E.P. Publishing for the Botanical Society of the British Isles, 1976

Pratt, A. *British Grasses, Sedges, Ferns and their Allies*, Frederick Warne, 1884

Roberts, R.H. 'The Killarney Fern *Trichomanes speciosum* in Wales', *Fern Gazette* 12 (1979), pp. 1–4

'*Polypodium macaronesicum* and *Polypodium australe*: a morphological comparison', *Fern Gazette* 12 (1980), pp. 69–74

Rymer, L. 'The History and Ethnobotany of Bracken', *Journal of the Linnean Society (Botany)* 73 (1976), 151–176

Scott, D.H. and Ingold, C.T. *Flowerless Plants*, 12th Edn., Adam and Charles Black, 1947

Shivas, M.G. 'The *polypodium vulgare* complex', *British Fern Gazette* 9

(1962), 65–70
'A Cytotaxanomic survey of the *Asplenium adiantum nigrum* complex', *British Fern Gazette* 9 (1969), pp. 68–80
'Names of hybrids in the *Polypodium vulgare* complex', *British Fern Gazette* 10 (1970), 208–209

Sowerby, J.E. *The Ferns of Great Britain*, J.E. Sowerby, 1855 and 1859

Sporne, K.R. *The morphology of Pteridophytes*, Hutchinson University Library, 1962

Stansfield, F.W. '*Asplenium trichomanes* and its varieties', *British Fern Gazette* 5 (1927), 77–79

Stansfield, F.W. 'Experiments in the propagation from spores of hybrid ferns, *British Fern Gazette* 6 (1931) pp. 86–8

Step, E. *Wayside and Woodland Ferns*, Frederick Warne, 1949

Stokoe, W.J. *The Observer's Book of British Ferns*, Frederick Warne, 1950

Synnott, D. and Baird, H. *Ferns of Ireland*, no.68, Irish Environment Library series, 1980

Taylor, P.G. *British Ferns and Mosses*, The Kew Series, Eyre and Spottiswoode, 1960

Wherry, E.T. *The Fern Guide* Doubleday, 1961

Index

BEYOND the RIVER

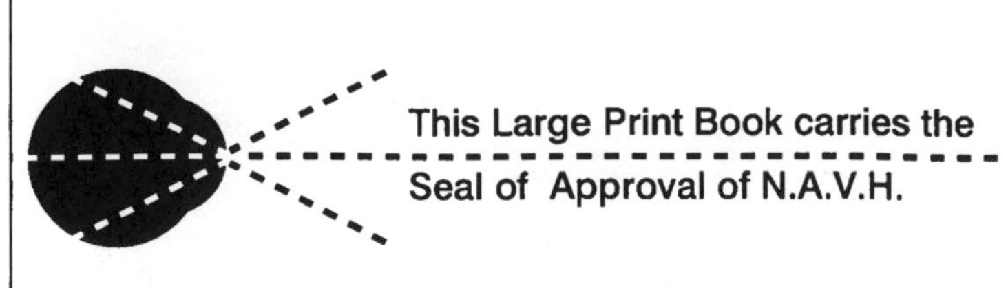
This Large Print Book carries the
Seal of Approval of N.A.V.H.

BEYOND the RIVER

THE FAR FIELDS SERIES
BOOK 1

Gilbert Morris

AND BOBBY FUNDERBURK

Thorndike Press • Thorndike, Maine

Published in 1999 by arrangement with
Starburst Publishers Inc.

We, The Publisher and Author, declare that to the best of our knowledge all material (quoted or not) contained herein is accurate, and we shall not be held liable for the same.

Thorndike Large Print ® Christian Fiction Series.

The tree indicium is a trademark of Thorndike Press.

The text of this Large Print edition is unabridged.
Other aspects of the book may vary from the original edition.

Set in 16 pt. Plantin by Rick Gundberg.

Printed in the United States on permanent paper.

Library of Congress Cataloging in Publication Data

Morris, Gilbert.
Beyond the river / Gilbert Morris and Bobby Funderburk.
p. cm. — (The far fields series ; 1)
ISBN 0-7862-1968-8 (lg. print : hc : alk. paper)
1. Large type books. I. Funderburk, Bobby, 1942– .
II. Title. III. Series.
[PS3563.O8742B49 1999]
813′.54—dc21 99-21835

Dedication

To *Denny* and *Dory Davis*
for
forty years of faithful friendship.

Contents

Part One:

THE CITY

Chapter One

The Trouble With George Washington — !

"I must be as crazy as they say I am — paying 500 Credits a month for this apartment!"

Starr had not spoken the words aloud, but as she awoke the thought came to her as it did almost every morning. A streak of irritation ran along her nerves, for she hated self-doubt, and as she sat up, she muttered, "It's my own business what I do with my Credits!"

Throwing off the light coverlet with more force than necessary, she sprang off the narrow bed, then stood there staring defiantly around her living quarters, which consisted of a single room, ten feet wide and twelve feet long. It was listed under Accommodations as LA-1, Luxurious Apartment, First Class. She had waited fifteen months for the privilege of paying 500 Credits each month, more than two-thirds of her total income. The walls were a neutral beige, as in all the other apartments, but the two oil paintings and the blue curtains on the single window added splashes of color. As she let her eyes rest on the one to her right, she steadfastly refused to think of what it had cost. It was very old, of course, coming out of a period which was her speciality — pre-NuAge history, but this painting went far back even for that period.

Now as she looked at it the strange sense of peace and serenity came to her as when she had first seen it. She remembered the moment when she had found it, on a visit to an antique shop in the Fringe. An old woman with blackened teeth had allowed her to go through the attic of her decaying shop, and she had discovered it sandwiched between stacks of yellowing newspapers and magazines. She had been searching for papers and periodicals, but when she had lifted an old magazine with the quaint title of LIFE and seen the oil painting, she seemed to enter another world.

It was a simple painting — a landscape with an OldAge farm in the foreground. A white house with a porch running the full width was in the center, and to the right a red barn with a pair of dappled horses. A huge oak tree stood on the left. A swing dangled from one of the massive limbs and two small children were being pushed by a woman wearing a long dress. To one side a man was smiling at the three as he stood holding the reins of a beautiful black horse. A blue sky, white clouds, and the rolling hills of green trees formed the setting. Something in the painting had moved Starr even then.

Looking at the painting she remembered how she had fought to keep it. She had bought it from the crone for only 50 Credits, but the Curator of the Past — a tall, thin man with narrowly spaced eyes — had given her a hard time. "It's not a *healthy* picture," he had argued. When she had asked why, he had snorted, "You're a Remedial Historian. If anyone should know the danger of the kind of rampant romanticism this painting glorifies, it ought to be you!" The Curator of the Past then gave her a stiff lecture on how Old World had perished because people wasted their time swinging children instead of serving the State — but in the end he had let her have the picture. He had let her have it for *only* 500 Credits, and Starr knew that most of that would go into his own account. But as she let her eyes rest on the painting and felt the sense of peace it always brought, she was glad she'd won that battle.

Starr glanced quickly at the other painting, one which her best friend had done. It was a strongly executed portrait of a young girl. Looking full face out of the frame, there was an air of innocence about the child that never failed to catch Starr's attention. One of the few people who had ever been in her apartment, a Clinical Dream Master named Lon Beta, had insisted, "You ought to get rid of that painting, Starr. It's a wish-fulfillment thing." He had grinned broadly at her, adding, "You're always wishing you could go back and be a little girl again. I think you've read too much about that old fable of Adam and Eve in the Garden of Eden." He sobered, gave his head a shake, then admonished her, "That was a nice story, Starr, but it's just a myth. Maybe all of us at one time or another would like to find a place where everything's perfect. But we never will. That's why we have people like me. If you want to dream about Eden — or a place like that old farm house in the other painting, I can help you."

Starr shook her head, "I don't need any of your little dreams, Lon. I like to look at things straight-on." He stared at her with a peculiar look in his grey eyes. "Everybody needs dreams, Starr. Life would be terrible if we didn't have dreams. I've seen a few people try to hold out, but they all came around. You will, too."

Turning from the paintings, Starr went to one of the two doors in the apartment which opened into the tiny bathroom. Stripping off her nightgown, she stepped into the shower, and for her allotted time let the warm water run down her body. When the water shut off, she longed, as always, that she could have a shower that did not shut off automatically; but as she dried and dressed she smiled, remembering how before she had her own place she'd shared a communal shower.

Starr walked to her kitchen which was a section of cabinets three feet wide on one end of the room. The designers had included a two-burner stove, a microwave oven, a tiny sink and dishwasher, a refrigerator no more than 18 inches square, and a cabinet for her food. Her friends thought she was insane to go to all the trouble of cooking, when pre-packaged meals were cheap, while *real* food was expensive and difficult to find. But as she dropped a piece of real bacon into the small frying pan she knew it was worth it. Starr enjoyed seeing the bacon curl as it fried, and smelling the rich robust aroma. She cracked the egg and carefully fried it in the grease, just enough so that the yellow would be nice and runny, then warmed a roll with sesame seeds on top. Finally she sat down at the tiny table with the meal before her, as she looked out the single window and past the structures of the skyline of the City. Always at that moment something came over her as she sat in her own place with the good food she'd prepared. Her study of Ancient Religions of America had revealed that people in OldTime had said "grace," a prayer of thanksgiving to their God. Starr had no superstitions about God, but there was something in that moment that caused her to think that she could understand those people who lived so long ago in such a simple time.

After making breakfast last as long as possible, Starr cleaned up the kitchen, then changed into her uniform. Standing before the full-length narrow mirror, she studied herself with a jaundiced eye. She saw a tall young woman wearing a light brown skirt and tunic with a dark brown leather belt. Over her breasts was a pair of crossed leather straps; a pair of calf-length nylon boots adorned her feet. She gave her slim figure a critical eye, deciding that she could use another two pounds. Then she leaned forward slightly, peering at her face. It was a squarish face with a strong jaw and high cheekbones. The complexion was smooth and the olive tone was not the result of a tanning booth but came from some distant aborigine as did the mass of auburn hair. A pair of steady grey eyes looked back at her. They were wide-spaced, deep-set, and shaded by heavy black lashes. It was not a beautiful face, but, she thought with satisfaction, it was durable, and would cover the strange passions that rose in her from time to time.

Suddenly, a fragment of poetry came into her mind like a small bird darting into a room though a window carelessly left open:

> There will be time, there will be time
> To prepare a face to meet the faces that
> you meet —

Her lips drew suddenly tight, almost hard, then she wheeled and left the room quickly. And, as always when she left, there was a small stab of fear. She had long known what it was, for she was a woman who had learned to know herself better than most. The apartment, she had long since realized, was her harbor. When she left it she left safety and peace behind. When the day was over she would scurry back with the same urgency that small furry animals seek their underground dens when snowflakes begin to fall and storm winds gather. Starr realized it was a weakness, but she could not change it. As she moved out of the apartment into the hustling world of the City she forced herself to stand straighter, putting on the face that others would see — one filled with a confidence that she did not feel.

A covey of people waited for the elevator, so Starr turned into the stairwell and walked down the nine stories to street level. She disliked crowds — which set her off from most people. Most of her friends had discovered that she liked to be alone — a habit that some of them found dangerous and anti-social. Most people, she had long ago discovered, were afraid to be alone with themselves, and found it necessary to be with others around the clock. This was not illogical, since most denizens of NuWorld were born in crowds, lived in crowds all their lives, finding isolation only in death — and even then their dust was mingled with the ashes of others in the Great Urn.

The hollow sound of her boot heels striking on the concrete stairs had a sepulchral quality that rang in her ears. When she reached the door a red light flashed and she spoke clearly, "Hello, Starr Omega." A hidden microphone picked up her voice, transformed it into an electric signal, then fed it into a computer bank. Her voice print was identified, approved, and returned to a speaker over the door which said, "Good morning, Starr. Have a wonderful experience today!"

The door slid open noiselessly, and as Starr passed through it she wondered if others felt as she did while going through the exit procedure. She accepted the fact that security was necessary to keep Undesirables and Barbarians out of the ordered life of the City. But somehow speaking into a machine made her feel like an idiot. And the fruity voice that answered her with a platitudinous phrase angered her. At times she had to restrain the impulse to say something obscene to the door, but that would mean getting her name on a list somewhere in the Department of Concern.

Stepping out of the building, she walked to one of the small vehicles marked CARRIER and got inside. The interior was all light tan plastic, including the vinyl seat which was no more than 40 inches wide, room for two people. There were no levers whatsoever for controlling the carrier, only ten small buttons and one large red button marked START. Starr pushed 37.04 and a 71.25, the latitude and longitude where the Department of History was located, and then the START button. The carrier moved off at once, its electric engine emitting a mild high-pitched whine. From time to time the guidance system made corrections, such as slowing to avoid another carrier or turning down another street. Most of the passengers Starr noticed in other carriers were either reading or staring blindly ahead.

Starr never read while riding, choosing to examine the scenery that floated by or the activity that went on daily in the City. More than once she looked up to where the huge geodesic Dome made of clear plastic covered that section of the City. She was always impressed by the scientific genius that had produced the Domes, for they were miracles of architecture. As she looked up to the peak nearly a quarter of a mile overhead, her practically flawless memory brought into her mind part of an essay she had read concerning the building of the Domes. It ran through her mind in orderly fashion, sentence by sentence, so that she could have read it and recited it as easily as she could have read from the original text. It was that gift which made her such a good historian, the ability to read and recall bits and pieces of information stored in the grey cells of her mind. As she noted the way the sunbeams were refracted by the interlocking triangular plates so that all of Section G was bathed in the smoky brown haze of a muddy light, she ran through the article in her mind:

According to the old legends, the world was destroyed by water in the distant past. While no one believes that myth in these days, we are conscious that the earth did suffer a catastrophe no less severe than Noah's flood. And it came not without prophets, for in OldAge many pointed out that unless man ceased to pollute the atmosphere he would destroy himself. But the system of OldWorld was built on commerce, and commerce ran on fossil fuels. Although the techniques of solar power, wind power, and other forms of power were known, there was no country willing to lead the world into a program of restoration reform.

So one day the world died.

Not as in the days of Noah, of course. Not in a seven day week. But there came a day when the atmosphere could take no more. The tiny bits of burned oil and other polluted atoms that had gathered in the skies and were held in place by gravity reached a saturation point.

So Planet Earth suffered the Greenhouse Effect.

This meant that the age-old biological and botanical processes simply malfunctioned. The earth was trapped inside a container of poisonous gas with no way to break out.

So men and women and children strangled to death, slowly and horribly, their lungs seared by the air which had become toxic. Once it had been miners who had died of black lung, but now the denizens of the planet were dropping dead by the millions.

The end of the world had come.

But man is a god. And a small group of scientists proved that man can save himself — and it was the Domes that proved to be the salvation of mankind.

In a mad race against death, which halted technological process in other areas, a few desperate scientists hit upon the concept of a hermetically-sealed Dome (which had been done on a small scale many times) and created gargantuan Domes. Inside these Domes the air could be purified. Plants could create carbon dioxide. Purifiers could create safe waters. The first Dome proved the theory was good, but it was only a small one, covering about twenty acres. But when it demonstrated that life could go on under glass, larger ones were built. The geodesic Dome is the simplest of all structures, and one of the strongest. Now the scientific community threw itself into a massive program which made the building of the pyramids look like childish toys! Soon entire towns could be contained in a single Dome, and even the largest cities on earth could be contained in ten or twelve Domes linked together by hermetically-sealed passages.

Starr's carrier pulled up to a towering windowless building made of high grade plastic. A metallic voice said, "Eleven Credits, please." Starr extended her left hand, palm-down into a slot and waited until she heard a tiny clicking sound, and when the metallic voice said, "Thank you. Be all you can today. You owe it to yourself," she removed her hand. The laser had read the tiny chip implanted in her left palm, then fed the information instantly to the Bank. On her next statement would be an item labeled "Carrier" with the date and time of the reading and the eleven Credit charge showing. Getting out of the carrier and entering the building she wondered idly what it must have been like in OldWorld, having to carry money. Or even trouble with the old credit card system. One of the first adjustments made in NuWorld was to issue every individual a number which was implanted in a tiny chip in the palm of his left hand. No need for cash then, every service or item could be instantly charged.

Starr entered the building, then spoke to the guard who nodded casually. Taking the elevator to the sixth floor she went at once to the Office of Remedial History. But R.H. was more than a single office; half of the entire floor was broken up into small cells for individual workers.

Her own Division, American History, was a cluster of these smaller offices, and she went at once to her Director, Emmett Tau, who was sitting behind a gray desk staring at the screen of a computer. In front of him was a console. From time to time his hands would flicker over it almost lovingly. Starr knew he was auditing the work of his staff. And she also knew that he was not finding any mistakes in their work, for he looked glum. Only when he caught one of them in a mistake did his eyes gleam with joy.

"May I talk to you before I go to work, Emmett?"

The Director looked up from the screen, his dark brown eyes fixed on her. "Go ahead." His eyes kept flickering back to the screen, but there was no way to stop him from doing that. *If a Roman orgy took place in my office* Emmett Tau was fond of saying, *I wouldn't even look up!* He was a man of one dimension, and his god was the work he did.

"My Mentor at the University says that I'm ready for the dissertation, Emmett. Will you allow me to take some time off to do the paper?"

Emmett tore his eyes from the screen and studied her. "What will your subject be?" he demanded. He had a harsh voice, surprisingly loud in a man so small. What he was really demanding, Starr understood, was to know if her research would benefit Emmett Tau and the American Section of the Remedial Division of the Department of History.

"I'll choose a subject that will be of use to us here in our Section," Starr nodded, putting a strong note of enthusiasm in her voice which she did not feel. Her plan was to do a paper on a subject that was based on English history during the Middle Ages of OldWorld, but it would be fatal to tell Emmett Tau such a thing. "I thought we might get together and you could help me choose a proper subject," she added demurely.

Emmett stared at her, his mind almost whirring as he thought how he could get every possible benefit from this situation. He noted the slim beauty of Starr Omega, thinking that maybe he would have to offer her a six month Loving Friend Contract. He added to her physical charms the bit of information that she was the best junior remedial historian he'd ever had. And it all added up.

"Why, I think we can work something out, Starr," he said. "Why don't we get together later this week and see what we can come up with?"

"Fine!" Starr smiled, then at once said hurriedly, "Hey, I'm going to be late!"

She turned and walked out of his office, going directly to her own, which was simply a space eight feet square with four walls and a ceiling. There was a hook on the back of the door for her coat. But the room had no other ornaments or decorations of any sort. Two pieces of furniture were located in the middle of the room — a small wooden desk and a straight back chair. On the desk were four monitors and one keyboard.

There were no books, maps, or charts in the

room. The small computer on her desk interfaced with the megaputor which occupied the first three floors of the building. There were no documents, magazines, or books in the entire Department, for all information was stored on tiny silicon chips. By simply touching the keys of the console Starr could bring up on the screen a copy of the Declaration of Independence, a letter from Lord Byron to his mistress, or the number of inches of rainfall in the ancient state of Nevada during the year 1969. Three of the monitors were for reading information from Megaputor; the fourth was for her own writing.

Since there were no windows or any other form of distraction, Starr went to work at once. That was what was intended, of course, and she didn't resent it. She was one of those rare individuals who could close out every outside stimulus so that her entire attention could be concentrated on a single task. Starr had read once that great hitters in baseball could do that. Many of her friends needed drug therapy to give them that sort of power. But that required the use of stronger and stronger drugs — which meant that ultimately they lost almost all power to concentrate.

Starr brought the computer to life with a touch of her hand, an act that made her feel like the huge figure of God on the ceiling of the Sistine Chapel. She enjoyed the power of having at her fingertips the history, achievements, and power of not only NuWorld, but of OldWorld as well.

For the next four hours she sat at the desk, pulling from the microscopic chips far below her feet, documents, letters, journals, maps, books, drawings and a host of other fragments of information. She saw them as part of a huge jigsaw puzzle, but much more difficult since most of the pieces she examined did not fit into her project and had to be discarded. That was what destroyed promising historians. They were not willing to wade through a thousand items to find one that would fit into the whole they were trying to construct.

The particular task that Starr was engaged in was to "remedy" a certain historical item — the religion of George Washington, the first President of America in OldWorld.

It was a problem more difficult than most, for the inhabitants of NuWorld were fascinated by OldWorld history. Naturally, for their roots were there, most of them. The Greenhouse Effect had one effect of its own, and that was to melt the polar caps. This meant that the waters of all oceans rose. And this meant no more Florida. No more Gulf Coast. No more Southern California. If an

OldWorld inhabitant could be raised and shown a map of NuWorld he would have had difficulty recognizing it. But changed as it was, it was still the homeland, and people wanted the story of OldWorld. Indeed, they never seemed to get enough of it!

But knowledge is power, and the new rulers of NuWorld had no intention of letting anyone get powerful enough to bring destruction on their new society. OldWorld had lacked control, and that had brought its doom.

And that was why *remedial* historians such as Starr Omega were needed.

It was easy enough to find out that certain events transpired in OldWorld, that certain men or women performed certain deeds. "That's no problem for a good historian," Emmett often told his staff, then his eyes would glow with a zealous light. "But history must be put in such a form that it will do no harm to those who read it."

Case in point — Mr. George Washington. He was the General of the Continental Army of the United States. Yes, no problem. First President of the new republic. Fine. Served two terms and died in the year 1799 (OldTime) Quite right.

But Washington had one tragic flaw. He was a man who believed in God. A man who prayed. It was difficult to put him into an ecclesiastical structure, for he did not lean to such things. Nevertheless, there was a mountain of evidence, sound historical evidence, to prove beyond a shadow of a doubt that George Washington did believe in God. And, unfortunately, did not keep this belief to himself. He displayed it often in full view of eye witnesses.

And now that NuWorld had come there was no need of God.

So the question that Starr had been struggling with was simple: How do we give the history of George Washington to the people without letting them know of this unfortunate blot on the great man's character?

That was the nature of a remedial historian. To take historical material, and when it conflicted with the philosophy or beliefs of NuWorld — to find a way to amend the material.

It would have been simple for Starr to simply ignore the facts of Washington's religious bent. But that was the method of a heavy-handed amateur. That method was used, but the problem with that approach was that it simply didn't work! Someone was always digging up information and revealing facts long after NuWorld remedial historians thought they were safely destroyed.

Starr was one of the new breed of remedial historian. She felt that the facts were sacred. But when they did not fit into the system, it was her job to find a new method of *presenting* them so that they would dovetail with "truth."

"It's a matter of selection and interpretation," she often said. "Facts are hard, but when they are placed in the crucible of selection and interpretation, they become flexible and then can be correctly grasped."

And it was her fiery and stubborn determination to face the whole "truth" that made her a valuable member of the Department.

"If Starr writes it up," Emmett Tau said firmly to his superior, "There won't be any skeletons in the closet for anyone to find. She'll find them herself and pick them apart bone by bone!"

But George Washington was not cooperating. Starr had doggedly pursued the man with every ounce of her intellect. She had examined literally thousands of pieces of information, some from his friends and many from his enemies. She had thrown upon his behavior and his life a harsh searchlight —

And had found nothing!

Finally, she reached out and switched her machines off, and then in a rare display of raw anger she struck the table with her fist.

"Blast you, George Washington!" she cried out, trembling with fatigue and anger. "You really *were* the greatest man of your day! And the whole thing would probably have fallen through if you hadn't held it together!"

Then she put both her fists on the table, leaned over and with her eyes closed, said in a hard, flat voice:

"And you *did* believe in God! And nothing I can say or do can change *that!* The trouble with you, George Washington, is that you didn't have the good sense to be an atheist! All you could do was save your country!"

Then she flung herself out of the room and left the building without speaking to another soul.

"What's wrong with Starr?" Emmett's assistant asked.

Emmett gave him a certain smile. "George Washington is giving her a bad time. But she'll get that rascal pinned down! You'll see!"

Chapter Two

The Reliever

Adolf Hitler was on the right track, Starr thought as she left the gleaming black obelisk, *but he had that single flaw that prevented him from fulfilling his potential. He was insane. Washington believed in God. Hitler was insane. Well, thank Freud I don't have to remediate Hitler! Some poor fraulein was probably fuming over that right now. How do you amend the attempted extermination of an entire race of people in the name of Christianity when the alleged Christ was a member of that race? Good luck Gretel or Heidi; you'll need it.*

Starr felt the hot anger flowing through her and knew it was counterproductive, so she did what she had been doing more frequently to control any outburst of emotion — thought of the painting in her apartment. On the front steps of her mind was the light. Had there really been light like that in OldAge or was it only the imagination of the artist? Those lovely shafts, like transparent gold from the sky, striking the leaves and exploding in green-gold starbursts! She looked toward the Dome far above and the muddy glow that seeped through it, bathing the City in a perpetual brownish haze. Surely there had been no world like that painting!

As Starr came back to herself she was walking on the narrow strip of concrete between the sidebelt and the carrier track. She looked around self-consciously and stepped onto the sidebelt. Directly in front of her was the Astral Mall with its massive rainbow-shaped entrance decorated with the twelve signs of the zodiac. She decided to grab a quick lunch before she made her afternoon house call.

As usual the mall was a sea of human bodies. The palm readers and fortune tellers were doing a thriving business. A barker in front of the Crystal Palace, dressed in a shimmering silver robe with the hood thrown back from his shaved head, was hawking business for his assortment of quartz crystals. "Kneel to your own self. Honor and worship your own being. God dwells within you as YOU!" He bowed and motioned her inside, but she shook her head and continued on. She had bought a crystal once but never could seem to use it correctly. One day she stepped on it in the dark and cut her foot. Then, with a vengeance she flung it down the disposal chute in her bathroom.

Starr stopped in front of the Dog Shoppe, which always amused her. In the display window the dogs were bouncing and chasing each other and barking their metallic little barks. Everyone thought it was a fad destined for the same fate as the Hovershoe and the Static Darmsbow, but the craze had lasted as long as Starr could remember. She stepped inside and put her hand over the rail into the play area and immediately the sensor of the closest dog picked it up. He was a Beagle, and his long ears flopped wildly as he bounded over to her and licked her finger with his wet and remarkably fleshlike tongue.

They were cute, she thought: Dalmatians, Shepherds, Chihuahuas, Great Danes, all four inches high and perfect in every detail. Real dogs, along with cats and most other household pets, had vanished of course — sometime during the Season of Black Skies when people were starving by the millions. She was tempted to buy one, but decided she must put away childish fancies if she were ever to achieve the proper Level of Ascended Consciousness.

The shopkeeper, wearing a tan smock and an unbearably pleasant smile, approached her from behind the counter. "May I interest you in one of our cuddly little darlings? The

Beagles are on special today." He reached down and scratched the miniscule dog on the back and its right hind leg became a frenzy of motion. "The batteries have a lifetime guarantee and there's absolutely no maintenance."

Starr looked at the Beagle, its hind leg slowing down as the scratching stopped. "No thank you. I'm out a lot. I'm afraid he'd only get lonely."

The shopkeeper frowned and shrugged his shoulders. "Strange times we're living in," he said, returning to his counter.

As she left the shop, she glanced at the logo above the door, a rainbow ribbon of light containing the words, "God is dog spelled backwards. Dog be with you."

Like most in the City, the restaurant was a typical fast-food affair designed to provide you with the proper nutrition in the shortest possible time. The food was virtually tasteless, as most food was everywhere, but no one seemed to mind. Soon she would treat herself to a meal in one of the run-down cafes in the Fringe, which was the only place left that she knew of where food could be consumed for pleasure as well as energy.

What Starr liked about this one was the loft area that was accessible by a spiral staircase. There was seldom anyone there because of the wasted effort of climbing the stairs. As she climbed, Starr looked out over the fifty or so clear plastic tables and booths, all filled to capacity. The men and women were dressed in the same type of uniform that she wore; only the colors varied, reflecting occupation, and rank was indicated by symbols sewn on the right shoulder.

Starr sat at one of the three tables that looked out over the shoppers, bustling along the slate-covered floor of the mall. A series of fountains ran down the center as far as she could see and people were sitting on the circular benches that surrounded them. They were lighted by recessed florescent tubing, powered by a system of electric pumps tied into the massive dynamos in the bowels of the City and covered by plastic bubbles to keep the water secure. She tried to imagine children playing among the crystal and chrome and black plastic columns of the mall. She couldn't.

A slot was positioned under the right edge of the table and Starr slipped her hand in, palm down. A lighted menu appeared in the table top. She touched "Salad and two smaller adjacent squares for extra vitamin C and extra iron." Almost immediately she heard a whirring noise in the black column next to the table, a panel clicked open and she reached in the opening and took out a tan plastic container and spoon. She took the cover off and looked at the six green globs of paste-like substance arranged in a circle on the plate. *They've hired a chef from the City's Culinary Department,* she thought, and replaced the cover. *Well, at least there's the loft and the walls don't talk to you.*

"Greetings Starr Omega. Power and life to you." A small young man with a terrible complexion and sad brown eyes walked up, grinning at her. He worked in one of the offices in her building, and pestered her whenever he happened to see her in the restaurant.

"No need to be so formal. We're all merely citizens under the Dome." She had no sooner said it than one of the Manuals in his slate-gray overalls flicked by her line of vision and disappeared into the nearly invisible hinged panel in the far wall.

"Hey, Starr, what about it? You ready to get with me? My Theosophy professor says a hands-on experience with a Reliever is a must for anyone wanting a well-rounded education. I'm even thinking of majoring in Relief." He stood almost at attention in his tan office fatigues with a hopeful expression on his face.

She wished she could remember his name. "Pimples" was all that ever came to mind.

"I'm afraid getting clearance from the proper authorities would be next to impossible in a situation like this. The Department of Extreme Concern has to approve them all now."

The young man paled under his acne. "Well, gee, it's not all that important I guess. There's always other areas of interest. Good-bye Starr Omega." He spun around, took the stairs two at a time and walked briskly across the restaurant and into the office.

Starr sat at her table, watching the masses moving below in their uniforms of muted shades of green and gray and brown. She sat and thought of the painting and of the light and the marvelous warmth and color of it, and she felt the soft breeze that stirred the leaves and set the light rippling and shimmering in the trees. *What in Freud's name am I doing?* she almost said aloud. *I have those pitiful citizens waiting for Relief and here I sit, lost in a dream of a forgotten world! You'd better start taking your responsibilities more seriously, Starr Omega! People are depending on you to fulfill dreams and lift burdens.*

As the harsh glare of the hidden florescent tubes drew her to the present, Starr lifted the plastic container and pushed a red button on the table's edge. A panel zipped open in the column, she tossed the container in, then stood up to leave. As she took the first step, a voice tinkled from underneath the table, "Thank you very much, Starr Omega. Please visit us again soon." A look of disdain came over her face. "Well, scratch another restaurant off the list," she thought as she headed for the stairs.

The carrier came to a whining stop in front of the Department of Adjustment. It was closer to the Rim than the Department of History and the Astral Mall. Consequently, the buildings were somewhat shabbier and there were scraps of paper about. Starr even saw a plastic spoon lying in one of the narrow alleyways. Rumor had it that some of the undesirables who had eluded Peacemaker Units were seen on these very streets in the early morning hours.

Above the main entrance, in block letters on the building's gray plastic exterior were the words, "To Protect and Serve." Starr entered the main lobby and saw a crowd around the bank of elevators. She disliked the vacuum tubes, but since her daydreams had put her behind schedule she turned right down a narrow hall and stopped near the end. She placed her palm on the white circle on the wall and the panel whooshed open. Stepping into the narrow cylinder, she sat in the contoured seat, leaned back into the headrest and folded the curved and padded bars around her. "Sixty-seven, please," she said and closed her eyes. There was eight seconds of high-pitched whining and a heavy weight pressing down on her.

As Starr left the tube she heard a familiar and unwelcomed voice, "Hey, Starr. Come here a minute, will you." Sammy Chi was crossing the room toward her. She knew precisely what he was going to say.

"How about a three month Loving Friend contract? You'll only have half the fun, but it's better than nothing." His white teeth stood out against the dark face as he smiled broadly.

"You might as well ask for three years, Sammy," she called out over her shoulder as she continued toward her office. "You've got as much chance at that as you do three months, or three seconds."

Sammy's face reddened slightly, but he was not put off. Catching up with her, he took her by the arm and turned her around. "C'mon, Babe, Don't be so hard on yourself! Give it a chance." The smile had returned as he leaned on one of the many gray plastic desks in the common area, running his hand through his curly black hair.

He was interesting, but vain to a fault Starr

thought, and she didn't have time today for verbal fencing. "Is that a gray hair I see, Sammy?" she asked, glancing toward the dark mass of curls.

The smile vanished. "Where? Show me where, Starr," he said in a sudden panic, bending his head down. "Pull it out!"

Starr turned and walked away, and as she entered her office she saw Sammy hurrying into the dressing room. "That'll occupy him for ten minutes," she thought. "Then he can pester someone else."

Her office was much like the one at the Department of History, except there was a gray plastic table instead of a desk and she had only one monitor. The space was eight feet by ten feet. There was an extra chair in one corner and a black cabinet in the other.

Starr sat down and touched the keyboard. The screen came to life. *1107723-Southeast Sector-Unit 43-Tube 49-Rim. Religious Dissident. Attitude untenable. That means only one thing,* she thought. *1107723 has gathered some scraps of Christianity from Freud knows where and refuses to have them expunged from his psyche.* She had never relieved one, but she didn't relish the task for these "Neo-Crossbearers" were the only clients never given the mind soothing chemicals that made them pliant as putty. They must face the most staggering of the "Unknowns" totally unassisted.

How bits and pieces of this ancient religion kept turning up and why some would cling so tenaciously to them was an absolute mystery to Starr. All other religions were tolerated and even encouraged, yet every effort was still being made to eradicate this pernicious behavioral plague that had proven so destructive to civilization in OldAge. All the artifacts relating to it had been destroyed (but the scraps kept surfacing) or, according to some, were stored in secure vaults in the White Tower above the Department of Concern. That, however, was only conjecture and she gave it little credibility.

Starr remembered the words of Freud, who loathed all religions, but, even in NuAge, had maintained his preeminence as the god of Behavioral History. *The whole thing is so patently infantile, so foreign to reality, that to anyone with a friendly attitude to humanity it is painful to think that the great majority of mortals will never be able to rise above this view of life.*

Starr knew almost nothing of Christianity. She had dumped it into her "dustbin for religions" along with all the others she had been exposed to. To her, they were all the same. She had the mind of Goethe, who said, "He who possesses science and art also has religion; but he who possesses neither of those two, let him have religion!"

"Well, to work," she murmured, summoning her resolve. She pressed a button in the right edge of the table and waited forty-five seconds. There was one knock on the door and a stentorian voice from the other side announced, "Peacemaker delivery for Starr Omega."

Why must they be so loud? Are they all required to attend Bellowing School? "Come in," she said.

The Exactor was about six feet tall and dressed in a uniform similar to Starr's, except that it was white with white nylon cross straps and white boots. Like all Peacemakers, he wore a white helmet with a mirrored visor that was always pulled down over his face. "Delivery accomplished," he said in a flat metallic voice.

Placing his hand palm down on the white circle on the top of the table the door opened. Then, he slammed the door as he left the room.

The man he had brought with him was dressed in the ubiquitous slate-gray overalls of the Manuals. He was about five-ten and bald, except for a gray fringe around his temples and down the back of his head and neck. He had pale blue eyes, thin lips and a nose that had been broken sometime in the past. His age was difficult to guess, maybe fifty-five. He weighed about one hundred and sixty pounds (the Metric System had never been accepted, even in NuWorld). His hands were strapped to a plastic transport belt around his waist.

Starr thought of him as harmless a person as she had ever seen. She was required to offer the dissidents who were not relocated a final opportunity to recant their beliefs, but she was taken aback by this one and was trying to formulate an opening statement. She had expected terror, not serenity. This, like all inappropriate behavior, was disturbing to her.

"My name is Philemon," he said.

"Philemon! No one is named Philemon," Starr replied without thinking.

"I am," the man said in a quiet steady voice.

Starr looked at him and tried to regain her composure. *Why is his presence so unsettling,* she thought. *I'd rather they came in screaming and thrashing about!*

"Aren't you supposed to talk to me?" the man asked.

After a thoughtful pause, Starr gave him the required appeal: "You can still be of service to the City," then looked at him closely. "What did you say your name was?"

"Philemon."

"Yes, Philemon. As I was saying, your situation is not hopeless. You need only denounce this insidious philosophy that has seduced you and the City will forgive your rebellion, embrace you and nurture you."

The man looked at her with those pale weary eyes that held no fear, that were filled with a kindly light and something else. "I'm a simple man," he said. "I don't understand your words."

Starr still couldn't place it as she looked at him. "I shouldn't tell you this, but all you have to do is make a statement. It doesn't have to change what you believe. Just don't talk about your beliefs."

The man smiled and Starr felt something. *Assurance, that's it,* she thought. *That's what he has. He's so absolutely sure of something.*

"If I don't share what I have, I lose it," he said. "Let me share with you the. . . ."

"Stop!" Starr said. "I won't hear this. I offer you life and you throw it back in my face. Enough." *Someone must have given him a drug somewhere along the circuitous route they all take, to make him easier to handle. He's not rational.*

"I'll pray for you," Philemon said softly, but Starr was no longer listening to him.

I must learn to control myself, she thought.

This is worse than that George Washington business! She touched the proper sequence of numbers on the cabinet, opened the heavy door and removed a white box the size of a man's heart. She placed the contents on the table: two pressurized syringes four inches long and the diameter of a pencil, an alcohol swab in a foil package (Why make it sterile?), and a two-inch by three-inch booklet entitled "The City — Departure Message."

Philemon's head was bowed and he was praying. All Starr saw was a man in extremis babbling to himself.

"I'm required to read this to you," Starr declared. The man raised his head and she saw that the assurance was unchanged. Starr read the brief message, ripped the foil package open and swabbed his left arm above the elbow. She held the first syringe to the sterilized area and pushed the button at the end. There was a pop like a distant air rifle and, as Starr turned away, she noticed Philemon's pupil's were normal.

Starr sat in her chair with her eyes closed. *I won't think about this,* she said to herself. *I'm a member of an honorable profession and the City has trained me well for it. I provide a valuable and indispensible service to our citizens.* And after a few seconds, *He won't be any problem now.*

The drug had taken Philemon out of the room, the City — and the world. He sat limp and docile. Starr took a pair of heavy scissors from a drawer in the table and cut the plastic thongs on his wrists. Then she had him stand, removed the plastic transport belt and laid it on the table next to the white box. Taking the second syringe and setting the timer for ten minutes, she tucked it inside her tunic and said, "Follow me."

Starr led the way down a dark narrow corridor and entered the third door on the right. Normally the entire procedure was accomplished here, but none of the standard rules applied for dissidents. The room was dim with a soft green light emanating from the top of the opposite wall. To the left was a reclining chair covered in a dark green fabric.

"Sit down, please," Starr said, and Philemon lay back in the plush chair.

Starr touched a sensor on the arm of the chair and the entire wall in front of it sprang to life. Even after witnessing it dozens of times, Starr felt that she was actually rushing at incredible speed through deep space. Giant asteroids zipped by and the entire solar system could be seen at a great distance with the sun's overpowering brightness and the steady shine of the planets.

Looking away from the screen, Starr took

the syringe from her tunic and walked to the opposite side of the chair. She looked down at Philemon. His eyes were closed and his lips were moving. *Poor thing,* she thought, *He's babbling to himself.* She placed the syringe against his left arm and pressed the button. There was a soft pop as the air pressure sent the clear liquid into Philemon's veins. There was nothing left to do.

It's gone extremely well, Starr thought as she left the room. *No complications whatsoever.* She closed the door behind her.

Two floors down, Lido, a Manual assigned to housekeeping duties, was startled by the sound of the slamming door. "No one closes doors that hard," he muttered and continued on his way in one of the service shafts that honeycombed the building.

Starr entered the break room and walked directly to the coffee service built into the wall. Taking a tan cup from the shelf, she held it under the spigot and touched the "Black-extra dark" sensor. The steaming liquid filled the cup one quarter inch from the rim. The room was empty and she sat down in a hard plastic chair next to the coffee service and closed her eyes with her head resting against the wall.

"Maybe I came on a little strong, Starrkins. You think we could start all over?" Starr groaned inside at the sound of Sammy's voice. *Good Freud! Will this man never learn!* She didn't respond in any way, merely sat there with her eyes closed holding her cup of coffee.

"Starr, are you asleep? Did you hear me?"

"Unfortunately I did, Sammy," she replied, keeping her eyes closed. "And to answer your questions chronologically: No-no-yes."

"I don't understand," Sammy said with a puzzled look on his face.

"I believe that, Sammy. I truly do."

Sammy sat in the chair next to Starr. "Is there something wrong with me?" he asked in a hurt voice.

Starr thought about Sammy and men in general. Maybe there was something wrong with her. She had no inclination to establish a relationship with anyone. "No, Sammy. There's nothing wrong with you. Not even any gray hair." She heard him laugh.

"That was a good one. Kept me looking in that mirror for ten minutes. How about going to the Sensory with me tonight? I hear they've just about got taste perfected in this new one. The other four have always been good, but I'd just about as soon have the OldAge movies the way this taste thing was going for awhile there."

Starr heard the drone of his voice, but none

of his words. She was thinking of assurance; pure, fearless, perfect assurance. And she saw a light in pale blue eyes that was somewhere still shining.

Chapter Three

The Far Fields

Bernard Alpha's office was larger than most. As director of the Department of Adjustment, with its four separate bureaus: Geriatric, Infant, Disease and Dissident, he was allowed two hundred square feet of space. He also had the only window in the building and it was huge by NuWorld standards, four feet tall and spanning the entire outside wall of the office, a distance of sixteen feet. Inside it looked like the rest of the tan wall. When he touched a sensor on his desk, it became transparent, but the dull black surface outside, which was kept polished to a sheen before the shortage of Manuals, remained unchanged.

The window was a status indicator with little practical purpose. The view it provided was of the blank wall of the adjacent building only yards away. As space was at a premium under the Dome, growth was vertical rather than horizontal and the proximity of the buildings made sweeping vistas impossible.

Bernard could, by standing at the extreme left of his window, look up and to the right, and see the slow curving fall of the Dome toward the Rim. On a particularly bright day, there seemed to be a slight tint of gold to the dirty brown light that filtered through.

Bernard touched his keyboard and Starr Omega appeared on his monitor. "Starr, would you come to my office please. I have an unscheduled duty for you."

Starr was leaving her office with the white box and the transport belt. She looked back at her monitor. "Yes sir. I'll just drop these off at Supply."

I hate to assign this unpleasant task to her, Bernard thought. *But she is competent and hasn't had field experience yet.*

Bernard opened the right bottom drawer of his desk and took an eight by ten black and white print from a hidden slot. *It wouldn't do for a man in my position to evidence any sentiment for OldAge.* He always used the print to calm him before any disagreeable chore, and during his lunch hour he locked his door and lost himself in it while he ate.

The photograph showed a park with acres and acres of trees surrounded by office buildings. There was a stream flowing through the open meadows and woods and people walking on footpaths and over bridges that crossed the stream. Boys and girls played on swings and slides and merry-go-rounds and families were picnicking on blankets spread in the shade of the trees. All this in the very heart of the city!

There was a single knock on the door of the office. “Come in,” Bernard said, hastily replacing the picture.

Starr entered and took one of the two hard green straight-backed chairs in front of the desk. “You wanted to see me, sir?” She liked Bernard Alpha. There wasn’t a hair on his head. His eyes were fringed with blonde lashes and brows and he was the most pleasant man she had ever known, in a position of authority.

Bernard lifted his five-feet-two frame from the chair and sat on the desk next to Starr. “Somcone has to make a house call and I’m afraid you’re the only one available,” he said in his soft clear voice.

Starr frowned and replied, “I’ve never been in the field before. You’re sending me alone?”

There was a chuckle from Bernard. “Don’t look so anxious, Starr Omega. I do believe you’ve heard too many wild stories in the break room.”

“There’s a rumor going around the office that someone has to make a ‘house call’ to the Rim today.”

"It's not a rumor. It's also not dangerous or I wouldn't be sending you, or anyone for that matter, alone."

"Going alone to the Rim doesn't concern me. I visit the restaurants in the area once in a while," Starr said, crossing her legs and leaning back in the chair.

Bernard placed both hands on the edge of the desk and leaned forward. "What's the problem then?"

Starr looked into his kind eyes. He truly was a different kind of bureaucrat. "Just a feeling I have."

"I watched you with the Neo-Crossbearer today. He got to you, didn't he?"

"It went perfectly," she said, unable to meet his eyes. She noticed the broken nail on her left ring finger.

"That's not what I asked you."

She was silent. *Must have happened during the Relief Procedure,* she thought.

Bernard had returned to his chair. "You haven't answered my question, Starr."

"I'm sorry," she said, looking up.

"Never mind," Bernard said. He touched a sensor on his desk and swiveled his chair half around.

Starr was watching the wall become transparent. "I didn't hear the question."

Bernard didn't repeat the question. He had

seen this happen before. "I was monitoring your Relief Process today, Starr. I trust you give no credence to the OldAge concept of death. That should have surfaced in the training period."

"I have no such illusion," Starr said formally.

"How do you view it then?"

"The same way any rational person would." Starr looked at the back of Bernard's head, then beyond him to the dark wall of the next building. "There is no such thing as death."

"Please continue," Bernard said as he swung his chair around to face her.

The face that Starr looked into was becoming harder. "There is only an endless recycling of the soul into body after reincarnated body," she said, returning her gaze to the broken nail.

"Exactly! And since there is no death, it only follows that there are no victims," Bernard said, not believing a single word he spoke. Beliefs didn't matter, or perhaps he no longer had any beyond his own immediate circumstances.

"It should have been different," Starr said. "I didn't expect him to act that way."

"Even in this age of enlightenment, human behavior is sometimes unpredictable, Starr," Bernard said.

"Yes," she said, unable to look up. "I understand that. It's only that he was so like . . . like a child. That's it. It was like working in the Children's Bureau, except he was a man."

"You need to reacquaint yourself with the basic precept of the Department, Starr," Bernard said, worried that she would not meet his eyes. "Do you recall it?"

Starr knew she must regain control of herself. "Yes, sir," she said sitting up straight and meeting his eyes.

"I'd like to hear it, please."

Starr recited "All our clients are no longer *karmically* in balance with the greater society. Therefore, we are rendering them the greatest possible service by enabling them to enter the next life."

"Excellent," Bernard said. "It is imperative that we free ourselves of the atavistic emotional baggage of OldAge if we are to cleanse the Earth and bring about the fullness of the New World Order."

Starr was in control again and looked coolly into Bernard's eyes. "I'm honored to be an integral part of the *Purification Process,*" Starr said, feeling her confidence returning with the words.

Bernard smiled and took a "field kit" from his desk drawer. It was a slim white box with the logo of the Department of Adjustment on one side and a clip on the other. "The address is inside," he said. "I'm sure you'll enjoy the trip."

"Thank you, sir," Starr said. She clipped the box to her belt and left the office.

After Starr was gone, Bernard considered the effect that such a brief experience with the man named Philemon had had on her. He had encountered only a handful of these "Crossbearers" and knew virtually nothing of their philosophy, as it was forbidden in NewAge. He believed them, as well as their predecessors in OldAge, to be a part of the ignorant and unwashed masses. That is until a few months ago.

He had been called to the White Tower on an urgent matter and, after its conclusion, visited an old friend who worked in the vaults. Quite by accident, he had been left alone with some of the artifacts of OldAge. Among them were portions of the writings of a man named Paul who had lived in the first century, by OldAge reckoning of time. He was evidently one of the exponents of this philosophy, and though he had had only a few moments to scan the writings, Bernard knew that this had been a brilliant and learned man.

Too many nights since then Bernard had lain awake with an unrelenting desire to read the complete works of this eloquent man who had written so long ago, of concepts unknown to him, of sin and grace and salvation, and of someone called Jesus.

Bernard Alpha stood at his window and looked down at the narrow carrier track and the base of the building opposite his. What he saw in his mind's eye was a color version of the black and white print in his desk. The people were moving about the park and he could hear the shouts of the children and smell the food on the tablecloths that were spread on the cool green grass in the shade of the trees. It was getting easier and easier for him to envision this scene. He felt that soon he would be able to join the people, eat with them, stretch out on the cool grass and fall into a long dreamless sleep.

The harsh buzzing of the monitor on his desk snatched Bernard from the vision. He was disturbed by the reality of his daydream. He crossed to his desk, but before answering the call he retrieved Starr Omega's personnel file from Central Records to his monitor, typed a brief entry into it and locked it in with the security code.

"How's my favorite Reliever today?" Ken Gamma asked Starr as she entered the station.

"Fine, Ken," Starr replied. "The sleds running on time today or should I catch a nap?"

"You got about ten minutes. Wanna go on up?"

She hated this part, another vacuum tube. But there was no sense in prolonging it. "Might as well," she said, enjoying the way he always catered to her with no demands attached.

Ken was on permanent duty at this station and he was her favorite. He had short brown hair and a round face and was always in a good mood. His uniform, which was a replica of the Peacemaker's, without the helmet and visor, was too tight for his pudgy body. For some reason this was not offensive to Starr, even though the NuAge ideal was the classic ectomorph body type. He reminded her of something she had run across during her research at the Office of Remedial History. It was in an OldAge children's book and was called a "Teddy Bear."

Starr sat on a bench and watched Ken go through the check list of the vacuum tube that would catapult her one hundred and fifty feet vertically to the loading platform that was a part of the Dome Sled track.

Stations for the Dome Sleds were located on the top floors of buildings that formed a geometric pattern throughout the city and at towers located around the Rim. Access, for the most part, was limited to official business. The sleds themselves were bullet-shaped stainless steel tubes containing five two-person benches, one behind the other. They had open windows that curved up into the roof and they ran on tracks that were attached to the underside of the Dome.

"All ready," Ken said as he bowed to her and made a gesture toward the tube with his left hand.

Starr took a deep breath and walked toward Ken. "I wish that tube wasn't clear. I'd rather not see where I am. Can't you paint it black or something?"

Ken chuckled. "Don't worry 'Starry Eyes,' I haven't lost a customer all day."

"I've got to get a job with more horizontal travel," Starr said to Ken as she stepped into the cramped chamber. Then there was a brief rush and the blur of the city all around her.

Standing on the steel eyrie of the loading platform, Starr looked down a thousand feet to the canyons between the buildings. Hundreds of carriers were sliding along their tracks like gray clots moving slowly and silently through the arteries of the City. To her left, the Dome rose slowly and inexorably toward the White Tower. It was a two hundred foot high white pyramid atop the thousand foot black tower that housed the Department of Concern. The entire structure had come to be called the White Tower and it stood in the exact center of the City.

At this height the light was much brighter as it came through the thick clear acryllic panels of the Dome, and it seemed to gather darkness as it descended. It glimmered and gleamed and moved in slow waves all around her and Starr felt she was in a city beneath the sea.

As she stood there gripping the cold steel of the hand railing, she thought of all the great cities of the world that had disappeared in the Season of Waters. *She saw an enormous shark, its eyes dead and cold as the depths of space, glide slowly through the murky waters of the Rotunda in the Capitol, a hundred feet beneath the surface of the Atlantic. Crabs were scuttling across the sandy bottom in the streets of Leningrad. She thought of New York and London and Tokyo, of Cairo and Barcelona and Melbourne, all taken by the sea.*

The Dome Sled creaked heavily into position at the loading platform. Starr remembered the words of a poet, or perhaps he was a visionary, from OldAge, *A savage servility slides by on grease.*

The long descent to the Rim began. Starr was enjoying the ride as she marvelled at the

Dome Sled System. It was built during the leanest period of the energy shortage, before the grain fields came into full production, and harked back to OldAge technology. An interconnecting series of cables and pulleys was designed to transfer the energy released by a descending sled to one that was ascending. This "gravitational" energy source had to be supplemented by electric power, but it was seventy-five percent self-sustaining.

Because of the nature of her "home visit," Starr was given an "Express Sled." As she passed the loading platforms along the track, she could hear bits and pieces of conversations:

"Peyote buttons? How in Freud's name did you manage that?"

"I have this friend who works at the Greenhouse and. . . ."

". . . at this precious little antique shop. It's an authentic Ouija from OldAge with. . . ."

"Holistic health shops are passe, my dear, and I'll tell. . . ."

". . . and having originated in the Lyra nebula, I'm able. . . ."

She soon tuned them out and heard only the low hum the sled made against the track and the occasional creaking as she shifted positions for a better view.

As far as she could see the stark windowless

buildings stretched in unbroken columns, following the invariable downward flow of the Dome. The government buildings were black, commerce and industry (which were government-controlled) were gray and the residential were tan. There was very little variation from this color scheme.

Ten feet above her Starr could see nests built on the surface of the Dome and birds of all sizes and colors were flying into and out of them carrying insects and worms for their young. Over the years debris, brought in by the winds and the birds, had collected in various areas and formed a kind of loose soil. To her amazement, Starr saw grasses and plants (and trees as tall as twenty feet high) were growing on the Dome. She was enthralled by the sight, for all the flora of the City was restricted to the government greenhouse and only authorized personnel were allowed admittance.

Starr tried to identify some of the birds and trees, but her knowledge was limited since there was virtually nothing on this subject in the general information pool. There was a plethora of files concerning the artifacts and mentifacts of OldAge, but information (especially visual) regarding its flora, fauna and terrain was practically non-existent.

A squirrel skittered across the smooth surface of the Dome and climbed a small tree that Starr believed was a variety of oak. He sat upright on the base of a limb and began eating a small nut of some kind. Starr could not identify it. She saw how the light hit him and gave off a reddish glow like something she had seen a long time ago but had no name for.

The sled was passing over the zone that marked the transition from Central City to the Rim. The tallest buildings here were three stories and there were no carrier tracks in the streets. But the streets were filled with people! This never failed to excite Starr and she loved to walk among them whenever she made a trip to one of the restaurants and she always came alone.

The black market flourished. You could buy jewelry and liquor and outlandishly colorful clothes of every design imaginable. There was a bustling barter system, and few of the people ever took part in the credit system of the City.

And entertainment! It was even more outlandish than the clothes. There were jugglers and acrobats and magicians on every street corner, or so it seemed. There were dancers and musicians on the streets and in the clubs, and plays were performed in the theatres. There were no Sensories. *Perhaps because of a lack of power,* Starr thought.

For years, the Peacemakers had tried to bring order and conformity to the Fringe, as this area was called, but it finally became a game of diminishing returns and they turned their efforts elsewhere. Anyone on the run from government authority was likely to be found here. Most of the stations along this section of the track were closed.

What Starr liked best were the artists. There were few studios. They mostly set their easels up on the streets and painted portraits and scenes of life in the area. Families gathered around their tables sharing meals, people dancing and drinking and celebrating at various events (it seemed they were always looking for a reason to celebrate).

One event in particular intrigued her. Someone had told her it was called a wedding. She had researched this and found it to be a custom of OldAge, but she was unable to grasp any real purpose for its continued existence.

It was a stimulating and mysterious area. And one mystery in particular attracted Starr — the paintings of one artist. He had a small shop near a restaurant she visited, but he would never speak of his work. "They're for sale, not for discussion," he had told her. There were scenes of vast plains and rolling hills and lakes and rivers. Some depicted a barren and desolate mountainous area. And some were of gold and green fields, broken by stands of huge trees that seemed to go on forever.

Starr had reached the final station. The heels of her nylon boots rang against the metal surface of the stairs as she descended into the very heart of the Rim. What struck her first was the laughter. Not the ringing, raucous laughter of the Fringe, but a warm, melodic sound that flowed over her like the showers she enjoyed so much.

People crowded the streets here, as in the Fringe, but they were dressed in the slate-gray overalls of the Manuals. She stopped two women as they walked by her.

"Pardon me. Could you tell me how to locate this address?" she asked, showing them the address she had taken from the box.

The women looked at the logo of the Department of Adjustment and their smiles vanished. "I'm sorry, I don't read," the older one said. "But you might check there."

"I'll read it . . ." Starr said, but the women had turned quickly and melted into the crowd.

Starr looked at the building the woman had pointed to. It was a long, gray, one-story affair constructed of sheet metal. There was a double line of these structures, all identical, curving along the base of the Dome as far as she could see. She opened a heavy iron door that swung easily and quietly on its hinges and looked into an enormous room filled with gleaming metal tables and chairs. The walls were lined with benches, and men and women sat and stood and walked about the room. Their expressions and actions seemed out of place in the cold stark environment of the great hall.

In a far corner of the room was a cubicle of transparent plexiglass panels. Inside it, Starr could see a row of gray metal filing cabinets, a desk with papers scattered on it and a lamp that glowed yellow in the white glare of the room. Above the door was a white panel with the word *Office* in black letters. A man with thin brown hair combed tightly to his skull sat at the desk, his head bowed over the papers.

As Starr crossed toward the office the people nodded and spoke to her. She returned their greetings, but felt uncomfortable doing so. She stepped into the open door frame of the office and the man looked up from his work. When he saw the emblem on the white box, he rose quickly from his chair.

"May I help you?" he asked.

Starr decided not to use the address. "I'm trying to locate a woman named Martha Epsilon."

"Yes, we were expecting you."

"You were?" Starr asked.

"Well, Martha was anyway."

"I see," Starr said, although she didn't.

"If you'll follow me, please," the man said as he walked around her and out of the office.

Starr followed him out a door directly opposite the one she had come in. It gave onto a narrow alley of bare concrete eight feet wide. The adjacent building had an open doorway and she could hear the sound of splashing water and the low rumble of a hundred separate conversations at one time.

The man turned left down another narrow alley between the communal bath houses. She followed him, her boots thudding on the smooth concrete. At the end of the alley, the man stopped and pointed to the right. Starr was unprepared for what lay at the end of the alleyway. She saw hundreds and hundreds of metal cylinders stacked at the very base of the Dome. They curved out of sight in both directions and were arranged in double rows of ten high and ten wide, back to back with the open ends out. They were four feet wide and eight feet long and appeared to have been welded together, forming single units of two hundred.

Men and women were going to and from the bath houses and climbing and descending the ladders that provided access to the cylinders. Most of them were barefoot. Some were in overalls and some had towels wrapped around them. All were dripping water and leaving wet footprints on the bare concrete floor. Starr looked to where the man was still pointing and saw the white panel with black lettering at the base of stack, *Southeast Sector–Unit 43.*

"Martha lives in tube seventy-three," the man said. "Will there be anything else?"

"No, thank you," Starr said as she stood in the controlled bedlam of the Rim, stunned, looking at the endless stacks of tubes. She noticed that the base of the Dome, to a height of fifty feet, was of opaque black acryllic and that, above it, the thickness of the panels and the angle made seeing through it impossible.

Starr followed the four foot wide passageway between the tubes. Some of them had cloths hanging over the ends; some were open. She noticed a pale woman in her mid-twenties sitting at the edge of tube number forty-nine. The woman had blonde hair and her light blue eyes were fixed on a picture she held in both hands.

Further along, to her right, Starr found tube number seventy-three, the home of Martha Epsilon. The lower rungs of the solid steel, forty foot high ladder were worn halfway through.

"Martha Epsilon?" Starr called, and waited.

Eight feet up a woman with short white hair that looked as if it had been cut with heavy shears looked over the edge of the tube. She had dark, olive skin and an open, unassuming expression on her face, which was surprisingly free of wrinkles. When she smiled, Starr noticed that most of her teeth were missing.

"Are you Martha Epsilon?" Starr asked.

"That's me, honey," the woman said. "Come on up."

Starr felt the smooth cool steel on her hands as she went up and thought of the untold climbs before hers that had worn the rungs of the ladder so deeply.

"Come on in. I'm making us some tea," the woman said.

There was a threadbare blue carpet on the floor with padding underneath to fill in the hollow of the curve. Three cushions lay along the right wall and two blankets, an extra pair of overalls and some odds and ends of clothes and other articles were stacked neatly at the rear. A cloth bag with a strap sewn on the side, packed as if for a trip, stood against the opposite wall. Near the clothes, on the bare metal floor was a small one burner stove. A copper kettle sat on the stove.

"Do you know why I'm here?" Starr asked.

The woman took two chipped plastic cups from behind her and placed them next to the stove. "Yes I do. You're going to take me to the Fields. That's why you're here."

Starr watched steam rise from the spout of the kettle as the woman poured the dark liquid into the cups. "I'm sorry there's no sugar. Times are hard right now," the woman said as she handed the cup to Starr. "Be careful. It's hot."

"I'm not here to take you to the Fields," Starr said. "I don't even know what you're talking about."

"The Fields," the woman said. "The beautiful green Fields where I was born. They're so far away I can't make it by myself. You've come to help me."

"I'm a Reliever," Starr said. "Do you know what that is?"

The woman looked puzzled. "Yes."

"I'm here to do my job."

"But I've done nothing wrong."

"I know that. I'll help you enter your next body. You're through with this one."

"This is the only one I'll ever have. When I leave this one I'll be pure spirit."

Starr saw there was no use talking to her. "If you'll lie down now, I'll help you."

The woman set her cup down next to the stove. "They came and took me away from my mama and my daddy when I was ten years old. I worked for them sixty-three years cleaning their buildings. Now I'm too old to work. Please let me see my home before I die, my beautiful Fields, my beautiful green Fields."

Starr had seen deranged people before. "Please lie down," she said.

"I will, child," Martha said. "I'm not afraid to die. I just thought I was going home first. I'm just an old woman, a silly old woman. I don't know why I thought that."

As Starr prepared the syringe, she looked at Martha. She was staring up at something that was not visible and she had a smile on her face that seemed to brighten the air around her. "I'm so tired and I'm going home. I'm finally going home."

Starr decided to use only one syringe. She used it and sat back and waited.

Martha lay back against the opposite wall with her head resting on the cold steel. Her coveralls were threadbare and the white socks on her feet made them look like a child's. A blue shawl, faded from countless washings, was draped around her thin shoulders. "How beautiful the heavens are," she said and lifted her arms upward. "Lord Jesus, receive my spirit."

Starr had never heard the name before and

the sound of it caused her to shiver. She was swept away by the immutable flow of eternity. Someone from long ago was calling her name and, in the echoing sound of a child's laughter, she was falling, falling into hands that were strong and gentle and sure.

Martha lowered her arms and turned her head toward Starr. "I forgive you, child," she said. "Lord, lay not this sin to her charge."

And when Martha had said this, her head moved slowly to the side and she fell asleep. Starr sat in the home of Martha Epsilon for a long time. And when she went down the ladder, she saw two men in white suits waiting at the end of the dark alley.

Chapter Four

Memories

Starr was trembling and couldn't stop. She stood in front of her bathroom mirror in the harsh glare of the neon tube. Her face was drenched with sweat and her hair curled in dark wet tendrils about her face, ears and neck. She touched the white dot on the counter, then splashed water on her face with both hands until it stopped running. Her face in the mirror looked strange, as though it belonged to someone she hadn't seen for a long time and couldn't quite remember.

The dream had come to her for the third time in as many nights, and the mood that descended on her afterwards was the same as last night and the night before that. No matter how hard she tried, she couldn't think her way out of it.

She stood in a small grove of trees and looked out across the Fields that stretched in an unbroken sweep of relentless green from horizon to horizon. At the top of the tree Starr stood under, a fox squirrel was eating an acorn and the tiny cuttings were falling on her like soft dry rain. The skies were dark and threatening; thunder rolled across the land like the sound of distant cannon. Far out in the field, an old woman with rough-cut gray hair, and wearing tattered overalls, was walking slowly away from her.

"Wait for me!" Starr cried out as she walked toward the old woman.

As the wind came, Starr began to run after the woman.

"Wait! Please don't leave me!"

She ran through the green, endless Fields as the wind grew stronger, flattening the Fields and tearing the tops of the trees. In the rushing of the wind was the name that the woman had spoken. The name was the same as the wind. When the woman spoke the name, the heavens opened and she left her body lying in the Fields and rose like a star into the bright open sky.

The wind subsided, along with the clouds. The woman also was gone, except for the flesh that she had been. Under a hard pale sky, Starr stood in the endless green Fields, realizing that now she was all alone in the earth.

Lights from the great height of the Dome that burned during all the hours of darkness gleamed softly on the blank walls of the buildings. They were purposed for combatting the crime problem, but Starr enjoyed the air of fantasy they lent the City. They softened the stark reality of day and created an atmosphere where she could escape into her mind.

The view from her window was sparse, her own block and parts of two others. It couldn't contain the worlds she imagined. But tonight, and for the last two nights, the dream had taken those worlds from her.

As she looked out she thought of an old newsreel she had seen in the course of her research — the Battle of Verdun in France in September of 1916. The constant shelling had blasted away all the trees; what remained was a moonscape of charred, twisted stumps, and trenches and bunkers packed with cold, weary, hungry men. That scene came to her tonight. She saw the desolate battlefield, with a dry wind blowing across barbed wire and shell craters; she saw torn bodies — and felt how very much like her soul this place was.

Starr stood next to Sammy Chi in a long line of people waiting to get into The Pentagram Club.

"It's the hottest thing in town, Babe," Sammy said as he moved to the rumble of the music coming from the building. "You're gonna love this place. I guarantee it."

Starr was familiar with Sammy's guarantees, but she was not going to stay home and wait for the dream. *I must really be desperate,* she thought as she watched Sammy gyrating next to her.

Inside the club the music was deafening. The rhythm was a frenetic hammering that threatened to crush Starr's eardrums, and if there was any melody at all, it lay crippled and helpless under the driving, pounding drums and percussion. Behind it all, Starr could hear what sounded like the wavering wail that came from minarets in ancient cities on the opposite side of the world.

"Whadda you think, Babe?" Sammy screamed into her ear. He had told her to dress in white, the "in" color for this week she supposed, and he had on a baggy white coat and pants covered in sequins.

He didn't want to hear her answer, Starr thought. She pointed to her ears, shrugged her shoulders and continued across the crowded club. She had on a white silk jump-suit with a red belt and high heeled red boots. She had bought them for the occasion, determined to liven up her life. The salesgirl had assured her the outfit was the perfect choice.

She noticed several men staring at her as she passed. She also saw the frown on Sammy's face. Her hair and nails were done to perfection. She had spent two hours on her make-up. Starr knew she was not beautiful, but tonight she presented a stunning appearance. Men were reacting as she had intended them to. But she was more depressed than she could ever remember being.

"Could we sit down, Sammy?" she asked.

Sammy looked across the sea of churning white-clothed bodies and then back to Starr. "Looks like we'll be standing for awhile, Babe," he screamed back at her.

Starr wished she had never told Sammy that "Babe" was sometimes used as a term of endearment in OldAge.

Purple, red and orange lights were flashing from the walls and ceiling. People were bumping and jostling them as they went to and from the dance floor. But there was no place left to stand. Starr wondered why the club was so full of smoke, until she saw thin streams of it rising from vents in the floor. *What's next,* she thought, *acid baths from the ceiling? What a choice I have. The dream or this nightmare.*

A waitress with a bald head and some fake horns came by carrying a red tray. On it were stacks of tiny white wafers. Sammy placed his right hand under the tray, took two wafers and handed one to Starr. Along with him, she placed the wafer on her tongue and felt it immediately dissolve. Starr felt herself moving away from her surroundings, but not far enough, she thought. She had heard that the drug's effect depended on the personality of the user. Sammy, she noticed, was jumping straight up and down like he was on a trampoline.

There was an electronic drumroll and the music and the people stopped. What was left was the wavering, high-pitched moaning that was grating on Starr's nerves. Curtains opened, revealing a small stage with a raised white altar. Starr knew what was coming.

A giant of a man in a leathery black suit that wrinkled and folded over his entire body walked heavily onto the stage. His head was covered by a square iron helmet that rested on his shoulders, and he wore massive iron boots that threatened to crash through the floor of the stage with each step.

As he raised his arms, Starr noticed that his hands were not visible under the suit. It covered the ends of his arms as it would the ends of posts or beams.

"Greetings from the Dark Kingdom," the man said in a voice that sounded like shattering glass. "Tonight, we shall awaken to higher realities and attain ascendent levels of consciousness."

"Let's go, Sammy," Starr yelled. "I've seen this all before."

Sammy looked at her with a puzzled expression, then his face brightened as he realized she didn't know. "No you haven't. The City passed an ordinance this morning. It's legal now. This is gonna be the real thing, Babe."

"I'm not up to it tonight," she said as she walked away.

A nude girl about eight years old was being led onto the stage by two black-haired women in white robes. She had long, tangled brown hair and her soft eyes were huge with astonishment, terror and betrayal. When she saw the man and the white altar, she began a hopeless struggle against the women who held her.

As Starr passed the giant in black, she looked up at the iron mask. Light seemed to be draining into the slits for the eyes and around the mouth ice crystals were forming as he breathed. She stopped and Sammy bumped her from behind. Starr walked on, thinking it was the drug that had caused this. As they left she heard the terrible screaming of the child.

Sitting next to Sammy in the carrier on the way home, Starr watched the couples strolling along the sidewalks. She hated the sidebelts and was thankful they were installed only in a small central section of the City. She enjoyed seeing the differing gaits of the men and women and always felt better about herself when she walked somewhere. It seemed such a healthy and wholesome thing to do.

Starr dreaded the inevitable confrontation with Sammy at the door of her apartment. *What a night,* she thought. *After this the dream can't be so bad. But it was.*

The door slammed loudly behind her as Starr stalked into her apartment. "That man has got to be part octopus," she said out loud. "Why do I do this to myself?"

She took off her boots and belt, then stripped off the silk jumpsuit, carried them into the bathroom, and dumped the whole pile down the trash chute. "That's much better," she said. "Never again, Sammy Chi. Never, never again!"

But she knew the problem was not Sammy Chi. He was no worse than most men she knew. In fact, he was probably a cut above the average, even if he did have eight arms. No, the problem was Starr Omega and she didn't know what to do about it.

Starr showered and put on her heavy cotton robe and thick cotton socks. Then she went into thc kitchen area and brewed a cup of herbal tea. Turning on her reading lamp, she stretched out in her recliner. She had paid a small fortune for the chair and knew it was more than worth it every time she ran her fingers over the soft brown leather. Taking her textbook, *Early Cinema in OldAge America,* from the table, she opened it to chapter twelve, *Romance — The Deadliest Form of Heart Disease.*

The text concerned itself primarily with the aftermath of romantic courtship: financial problems, unwanted and abused children, wife beating and divorce. There was precious little film footage in the archives, as anything other than interior scenes and selected shots of cities was prohibited, but Starr was fascinated by the ancient customs.

The bringing of flowers was her favorite. She often imagined some well-mannered (manners seemed important then) and gentle young man showing up at her door dressed in an old fashioned coat and tie and handing her a bouquet of red roses. She would put them in water in a porcelain vase. Then, the two of them would sit on a sofa in the parlor and talk. They would speak of poetry, music, and art, and of their families.

"And how are your mother and father?" he would say, taking her hand in his and smiling.

"Very well, thank you. And yours?"

"Fine, thank you," he would reply. "And your aunt Martha? I trust she's feeling better."

Starr tried to remember her mother and father, but their images were indistinct and so very far away. She seemed to recall a certain fragrance and a softness that was her mother, and the sound of her voice humming a simple sweet melody and the steady soothing motion of the rocking chair. Her father was gentle with strong hands lifting her higher and higher and swinging her around and around as she giggled and shrieked.

Nights in her bedroom were mommy and daddy kneeling by her as she lay snuggled in the cool sheets and warm blankets. Their voices were so beautiful and close as they talked — as they talked with someone they knew very well. Their love for her was like the warmth and softness of her bed. Her mother's song and her father's laugh folded around her and carried her safely through the long night.

The long night. Starr laid the text on the table next to her chair. What would carry her through this night — and the nights that lay before her like sworn enemies marshalling their dark forces? How would she ever endure those long, long nights to come?

There was a dreadful sobbing in the room and Starr sat up to realize it was coming from deep inside her chest. Tears were dropping slowly from her cheeks and sparkling in the light of the lamp as they fell on her trembling hands. After a time it stopped and she curled herself in the chair and waited for the dream.

The next morning Starr was at the stove

watching the bacon wrinkle and pop in the frying pan as she breathed in deeply the heady aroma. She thought again of her parents. Her Caretaker had told her that they had "reached untenable positions of social maladjustment." She was all too familiar with the consequences for those residents of the City who distinguished themselves in this fashion.

How different her life would have been had she remained with her parents? She was taken from them when she was five, only because they hid her. Children were routinely taken into the custody of the City when they were two, as parents were not deemed suitable to rear them. The Caretaker system had been in effect for generations and was regarded as the most effective means of producing responsible, conforming citizens. It was designed to mitigate, if not eliminate, counterproductive emotions such as tenderness, empathy and love.

It might be nice to visit them once in a while. Some of her friends did that, although not often, and most of them never mentioned their parents. The sound of her mother's song was coming to her as she forcibly thrust it from her mind. *Not now,* she thought. *This is not the time for that. This is the time:*

To prepare a face to meet the faces that
you meet —

She looked at the bacon. It was burned black and brittle. An acrid smoke was rising from the skillet. She scraped the bacon into the trash and tossed the skillet into the sink. Then she went into the bathroom, got into the shower and turned the water on as hot as she could stand it.

Sheila Phi paced her office and waited for Starr. She would stop occasionally and stare at her portrait of Gertrude Stein. She loved the strong German face and the thick beautiful hair and the lovely dark eyes. If she could have been anyone in history it would have been Gertrude Stein. Stein had not only a portion of fame, having associated with Hemingway and other noted writers of the 1920's, she also had a succession of nubile and talented young proteges who lived with her.

However, Sheila Phi had no talent for writing (Some would say, "Neither did Gertrude Stein.") and became a history professor instead, because of her remarkable memory and because she loved any era but the present.

As all offices are, this one was a cloudy reflection of its occupant. Other than the portrait, there was a small bookcase with volumes that were never opened (all her reading was done on the monitor), a hard straight-backed chair, and a desk with an artificial rose in a squat plastic vase.

Starr was late — this was out of character for her. Sheila Phi noted this and pulled up other relevant information from the storage banks in her brain. Lately Starr's appearance was somewhat unkempt. Her hair tended to disarray and there was seldom any makeup. The brow wrinkled frequently and there was a hint of dark circles under the eyes.

This information was assimilated and processed by the computer inside Sheila Phi's skull, and the printout contained one word — vulnerability.

Sheila Phi had been pressing Starr for a decision for months now and today could be the culmination of those bouts of verbal fencing. She had students more attractive and maybe one or two more intelligent, but there was something different (and she hated that word) about Starr Omega, something that refused to be constrained by the rigid environment she lived in. Today could be the day Sheila Phi got her decision about the *Loving Friends* contract. She looked once more at the broad peasant face of Gertrude Stein.

The Caretaker System provided the ideal soil for the growth of Sheila Phi's garden of dark flowers. She had been an unlovely child and developed cunning at an early age. She was short with stubby fingers and a tendency toward overweight which she learned to control as an adult. Her right eye had a cast to it and her nose was short and pointed and tilted downward. The widely spaced eyes were as vacant as a cat's, until she wanted something. Then they were whatever she told them to be.

Boys either avoided her or made fun of her and she came to regard men as emotional and intellectual inferiors. She made academics her god and all her pleasure depended on his benevolence. Relegated to the company of females, she developed a barter system for sexual favors. She planned to employ this system today as she sat in her hard chair behind the plastic rose.

This scene had its equivalent in the natural world in the way a hyena would patiently await the collapse of a young Thompson's Gazelle wounded by the lions.

Starr dreaded today's appointment with her advisor in the Department of History more than ever. Sheila Phi had been putting pressure on her for some time now to enter into a six month's contract with her, and she could make it extremely difficult for Starr to complete her dissertation if she were so inclined.

Starr thought of the mica glint in the eyes and the rigid, controlled movements of the body as she stood outside the door of the office. She knocked once.

Inside, Sheila Phi smelled the blood from the wounds of Starr Omega. "Come in," she said.

"Good morning, Omega," Sheila Phi said. "Have a seat."

"Thank you."

"You're not looking well. Something troubling you?"

"No, nothing," Starr said, avoiding those eyes that seemed to see inside her. "A little problem sleeping. That's all."

Sheila Phi got to her feet and walked slowly around the desk. She stood directly behind Starr's chair. "Sometimes loneliness siphons away our sleep," she said. "Sensitive girls like you need someone to share the long nights with."

Starr felt the hard stubby fingers pressing into the flesh of her shoulders and stroking the sides of her neck. Her stomach became queasy and her body involuntarily stiffened and pulled away. "That's not the problem. I still haven't come up with anything for my dissertation. Maybe that's it."

"Well, to business then," Sheila Phi said as she returned to her chair. "As we've discussed in the past, the relationship between student and advisor must of necessity be open and intimate to achieve maximum results. This is, of course, epitomized in the Loving Friends contract where symbiosis flourishes. The advisor as well as the student must grow to produce optimum gains."

"Yes, I see the wisdom of that," Starr said, "but my apartment is ideally located for school, as well as my work, and I was on that waiting list for such a long time."

Sheila Phi left her chair and began pacing the office floor. "Some things are more essential to one's career than geographic proximity, Omega." She stopped in front of Gertrude Stein's portrait.

"Perhaps you're right, but I'm sure that you're aware of the hardship this would entail."

Sheila Phi looked at Starr, then at the portrait. "What do you know of hardship, Omega?"

Starr assumed the question was academic and made no attempt to respond.

Sheila Phi was still looking at the portrait. "Gertrude Stein knew true hardship. Can you imagine the humiliation she suffered, an enlightened woman, in the age in which she lived?"

Starr remained silent.

"What a handsome woman," Sheila Phi said. "And such talent. Are you familiar with her work, Omega?"

Starr looked at the portrait and thought the expression accurately reflected Stein's writing, painfully constipated. "Somewhat," she said.

As Starr left Sheila Phi's office, she felt her life closing in on her. She had managed to postpone the Loving Friends contract temporarily, but eventually she would have to accept it or leave her studies, and the thought of either was unbearable.

She considered the events of the last few days and knew the stress was fast approaching her level of tolerance. *I've got to come up with something to get me out of this mess, if it's only for a few months,* she thought.

Chapter Five

A Way Out

Starr walked the streets of the City of her birth. It seemed more like a prison now and her mind was formulating escape plans. The people she passed all wore the same complacent masks. None of them seemed unhappy, nor did they seem happy. Why hadn't this seemed unusual to her before now? The City was able to pleasure all the senses of its residents and did so liberally. In doing so, it created a lukewarm population with no sense of hot or cold.

Sheila Phi's stubby fingers were again stroking her neck and Starr felt her stomach going queasy. Pushing the memory from her mind, she let her thoughts turn to her last two assignments with the Department of Adjustment. She tried to sever her emotional reactions and examine them in the light of scientific detachment. There was no logical explanation for the behavior of Martha Epsilon and the man named Philemon. No wonder the City found it necessary to “relieve” them. They didn’t respond normally to the Pleasure-Pain Principle that was the foundation of government.

After an hour, Starr found herself in front of the Department of Adjustment. She was not scheduled to work and decided to take advantage of the access the building afforded to the Dome Sled. A trip to the Fringe was just what she needed to get her mind off her problems. Maybe then she could get down to some rational, logic-based decision making. Starr thought of the one hundred and fifty-foot vacuum tube that would catapult her from the Dome Sled Station to the loading platform at the roof of the Dome and almost changed her mind — and the course the rest of her life would take.

Starr shared the Dome Sled with nine other passengers. As always, she was content to enjoy the dizzying view of the City. *It should be cold,* she thought. Having been raised in the climate-controlled Dome, Starr had never experienced true cold, but now the stark angles and shadows of the City somehow had the look of winter. She could almost feel the touch of an Arctic wind.

The woman next to her was in an expansive mood and insisted on sharing “her reality” with Starr. Ignoring her proved to be a fruitless endeavor.

"I simply must share this with someone," she said above the jangle of her bracelets and chains. "I'll just explode if I don't."

Starr took a deep breath and leaned back on the bench. There was an urgency in the abrupt movements and wide eyes of the woman, but it was the underlying pathos that stirred something inside of Starr and caused her to listen.

"Well, after going off on tangents for years, I decided to get back to basics — the suspended pyramid and twelve crystal circle," she began excitedly.

Starr looked at the other eight passengers in front of them. None took any notice of the woman.

"And of course the special quartz crystal, given to me by the 'enlightened master,' taped to my 'third eye,' " she continued, tapping the middle of her forehead with her index finger.

"Enlightened master?" Starr asked. Immersed in her studies for so many years, she had never paid much attention to the plethora of spiritualists and religions that came and went like the latest fashions in clothing.

"Oh yes, Lord Krishnamurti, of the 'Higher Council of Universal Masters.' He introduced me to the 'Keys of Enoch.' "

"I'm afraid I don't understand what you're

talking about," Starr replied.

The woman spread her arms wide, then dropped them and shook her head as if unable to measure the extent of Starr's ignorance. "It's gained favor again with all the enlightened," the woman said.

"What has?"

"The Keys of Enoch."

"I thought we were talking about the Universal Councils."

"Masters," the woman corrected. "Higher Council of Universal Masters."

Starr stared blankly at the woman.

"Surely you know they open the paths for us to be activated into a higher consciousness of light?"

"Ah — certainly," Starr replied, in the hope of thwarting another tirade by her unsolicited benefactor.

"Thank Freud!" the woman said. "I was beginning to think you were a complete spiritual illiterate."

"Hardly," Starr replied confidently. "What about these 'Keys of Eunuch?' "

"Enoch," the woman said. "Not these — it. It's a book."

"Oh yes," Starr said, trying to look interested. "I believe I remember it now."

"Well you certainly should," the woman said. "That is if you're interested in unlocking the mind-gates to the 'Spiritual Hierarchy of Light.' You are, aren't you?"

"Oh, yes," Starr replied, thinking that this trip was doing anything but clearing her mind.

"As you know then, the sixty-four keys to a complete cosmic synthesis of the spiritual and scientific realms will totally reconstitute the Earth and its peoples into a New World Order and. . . ."

"Excuse me," Starr said. "I thought we were talking about your getting back to basics. You know, the crystals."

"Yes, we were, but I get so excited at the thought of combining our New Age Spiritual Philosophy with Sacred Science to bring about the 'Second Genesis.' Just imagine, quantum physics, genetic engineering, holography, nuclear fusion technology and who knows what else, all perfectly meshed with the spiritual realm to create a Global Renaissance, a utopia."

Starr was totally confused, but she sensed that this woman was desperately seeking to fill some void in her life so she listened. Perhaps she also felt a certain kinship toward her.

"I can see you're a little confused," the woman continued. "So anyway, after I arranged the crystals, I sat directly beneath the pyramid — you know, in the 'crystal energy field.' "

Starr remained silent.

"You don't understand, do you?" the woman asked. "Let me refresh your memory. It amplifies 'higher vibrations' for receiving channeled thoughts from the spirit guides."

"Here's my station," Starr said, standing up. "I've certainly enjoyed our conversation."

"So have I. So have I. Maybe we could get together again. I'm being re-circuited into a whole new dimension of higher consciousness. You could join me."

"I'm afraid I'm a little busy these days," Starr replied as she stepped onto the platform. "Thanks anyway."

The woman's smile faded. "Surely," she said and looked directly into Starr's eyes as the sled pulled away.

Starr knew she would not soon forget the look in the woman's eyes and left the station in a gathering sadness.

The streets of the Fringe were literally bursting with the energy and excitement of the people. A small group to her left crowded around a bench where two men were seated playing music on trumpet and saxophone. Everyone was clapping their hands and moving with the rhythm of the music.

Starr tried to push the woman from the Dome Sled out of her mind. *Why can't people just enjoy things the way they are,* she thought?

Why are they always reaching for something beyond themselves?

As Starr stood there listening to the music, she noticed a boy of about ten tap dancing next to her. He had on shiny gold shoes, baggy purple pants and black shirt covered with gold sequins. When he saw he had gotten her attention, he stopped and bowed.

"You like my dancing?" he asked.

Starr was amused by his openness and innocence. "It's a thing of rare beauty," she replied. "You may very well be the next great star of the Sensories."

"Thank you, beautiful lady," he said, taking another bow. "You work for the government?"

"Yes I do," Starr replied. "How did you know? Are you a mind reader as well as a dancer?"

"I can just tell," he said. "I like your outfit. Especially those nylon boots."

"Why thank you. I like your clothes too."

"I bet I can tell you where you got those boots," the boy said as he began dancing again.

"I don't believe I understand you."

"It's simple. I can tell you where you got your boots. I can even tell you when you got them." The boy had stopped dancing and stood with his hands on his hips.

Starr was taken aback by his arrogant tone. "You most certainly cannot," she said.

"Wanna bet?"

Starr knew there were a hundred stores in the City that sold boots like hers. Even if he guessed the right one, he could never tell her when she got the boots.

"You scared, pretty lady?"

That was it! Starr prided herself on her intelligence and knew this child was no match for her, but decided that he needed to be taught a lesson. "No," she said. "I'm not."

The boy took a cheap gold bracelet from his pants pocket and held it out to her. "What you got to bet?" he asked. "Something good. This bracelet belonged to my grandmother."

Starr took a plain gold ring from her finger. It was very expensive, but she knew there was no chance of losing it. "How about this?" she asked.

The boy took it from her and examined it closely. "It'll do," he said, handing it back.

"You got your boots on your feet, on this street, right now," the boy said.

"What are you talking about?" Starr asked.

"I told you I could tell you where you got your boots and when," the boy replied.

"You made me believe you could tell me where I bought them," Starr protested.

"I can't make you believe anything, pretty

lady," the boy said. "You just didn't listen."

Starr realized she had been outsmarted. She looked at the boy smiling up at her, hand outstretched. "You'd have been a great politician in OldAge," she said.

The boy looked puzzled, then smiled again as she handed him the ring, bowed gracefully and walked off into the crowd.

"What's your name?" Starr shouted as an afterthought.

"Pan," the boy called back over his shoulder.

Starr listened to the music for awhile before she continued on into the Fringe. She walked past a fire-eater on the next corner, watched a knife thrower for a few minutes and stopped at a small three-wheeled cart where a man was selling hot dogs. She counted all the calories as he placed the hot wiener on the soft, warm bun; thought of the cholesterol that would clog her arteries as he spooned on chili, onions and mustard, and forgot everything but how wonderful it tasted with the first bite.

Starr sat on a bench as she ate, and watched the people go by. She marvelled at their energy and animation. They all seemed so excited to be about their business, so eager to get to wherever it was they were going. In the Caretaker System, they were all warned against the enticing lifestyle of the Fringe.

"They produce nothing, contribute nothing, think of nothing worthwhile," one of her teachers was fond of saying of the people who lived in the Fringe. They certainly seem to have a good time though, Starr mused.

After browsing through the shops and picking up a few trinkets, Starr found herself in front of the studio of the artist who was her personal favorite. She entered and saw him at work on a canvas in the rear of the small room.

"Well, well, if it isn't my favorite looker," he said, noticing her. "Make yourself at home."

Starr was watching the canvas come to life. "That's not fair, Jim. You know I'd buy all of your paintings I could afford if they weren't considered, ah. . . ."

"Subversive," he said, finishing the sentence for her. "We couldn't have that now, could we?"

"Exactly. The one I bought from the antique shop has put me under enough scrutiny."

Jim resumed his work while Starr watched, fascinated. A man stood at the edge of a wood under a huge tree that looked across an open field. In the distance, storm clouds were gathering in a dark swirling mass on the horizon, beginning to climb the sky. The wind was flattening the tall grass as it blew across the field toward the woods and the man, who looked inconsequential beneath the trees, appeared transfixed by the power and fury of the storm.

Some time later, Jim laid his brush down and washed his hands at a small sink. Starr hardly noticed him and continued to gaze at the painting, lost in the storm.

"How about some coffee?" Jim asked, pouring two cups at a table near the sink.

Starr came out of her trance. "Thank you," she replied. "It's a magnificent work."

"Hardly," Jim said, handing her the cup. "But it's not bad."

Starr looked back at the painting. "You really have some imagination," she said. Jim's face was toward her as he sipped his coffee, but his gaze had gone somewhere beyond her, beyond this room, beyond the City. Realization hit Starr like a physical impact. "You didn't create this scene!" she blurted out. "You painted it from memory!"

"I told you these paintings are for sale, not for discussion — even for exalted members of the hierarchy," he said absently, running his hands through the long straw-colored hair.

Starr looked into his stern face. "Have you really seen places like this?" she asked. "It's wondrous. What about the cloud of pollution that blocks out the sun? How did you get out of the City? Aren't the Primitives dangerous?"

The green eyes softened and Jim smiled. "My, my. Just look at you. I believe there's a woman's heart beating beneath those leather straps. Such passion."

Starr reddened and turned toward the painting. "It's just — just that I thought all of that was gone!"

Jim felt something come alive in Starr Omega that caused him to trust her. "Wait here," he said.

She watched him move to the front of the shop. He put the Closed sign on the door, locked it and pulled the blinds. Then he returned and arranged two chairs next to a floor lamp.

"Why the secrecy?" Starr asked after they were seated.

Jim leaned forward, his elbows on the arms of the chair, and rested his chin on his folded hands. "The earth is cleansing itself, healing itself, Ms. Starr Omega. That's the big secret. No one but the top echelon of our benevolent leaders and the Border Guards, of course, are supposed to know this most marvelous of all secrets."

"But the air — the water, they're poisonous!" Starr was incredulous. "The life span of the Primitives is only half of ours!"

"The air is pure. The water is clear and sparkling again," Jim said, leaning back with his arms resting loosely on the chair. "And the Primitives — only the strongest survived those terrible early years when everything you touched was poison. Now they're healthier and stronger than the best of us."

"Are you absolutely sure of this?" Starr was out of her chair, pacing back and forth.

Jim looked up at her and smiled again. "Omega, my dear, the greening of the Earth has begun."

"But why haven't we been told? Why are we still forbidden to leave the Dome?" Starr sat down, her mind reeling from this staggering knowledge. "It must be for our own protection. The criminals that have been banished would be a threat to our lives."

Jim looked at her evenly, without replying.

"That's it, isn't it? It's for our own good," Starr went on, not believing her words even as she spoke them.

"This is the place of death, the City," Jim said. "In the Fields, there's life."

"I can't believe that. I can't."

Jim reached over and took her hand. "The greening of the Earth has begun, Starr. Believe it."

"But why? Why would they hide this from the people? Why keep us inside the Dome away from the beauty and — and the freedom of the outside world? It doesn't make sense."

"History is your discipline, isn't it?"

"Yes," Starr answered. "Tell me why the Berlin Wall was built?"

Starr knew Jim didn't expect an answer. It was the key issue at the heart of every government. Freedom!

The revelation Starr had undergone was overwhelming to her. All her life she had believed everything outside the Dome to be alien and dangerous, placing her faith only in the City for safety. The City had been her refuge, her provider, her family since she had lost her parents. Now this world was gone, shattered, and she felt only loss and betrayal. How could she ever trust again?

"Are you all right?" Jim asked.

His voice seemed to come from a great distance. She struggled from the fog that seemed to surround her. "Why did you tell me this? Why trust me?"

"I can't say for sure. You've been coming here for a good while, but this time something was different — something I can't put into words. I felt you needed to know this." Jim sipped his coffee and smiled at her. "Besides, if you tell anyone, we're both goners. You know that better than I do."

In her confusion, Starr's thoughts returned

to Martha Epsilon. "I met a woman," she said. She couldn't bring herself to speak of the circumstances of the meeting. "She was very old and told me she was born in the Fields. She came to the City as a child, but all she could talk about were her beautiful green Fields. She wanted to go back there. Leave the City and go back there to die. What's so special about the place — that she would want to return after all those years?"

"I'm a painter, not a poet, Starr," Jim replied with that far off look in his eyes. "Maybe no one's words would be enough. It goes beyond the beauty of the land."

"There was something else about her, Jim."

He looked at her with a warmth in his eyes Starr had never seen before.

"She spoke of how beautiful the heavens were and she said a name — a name I've never heard before." Starr felt herself being drawn into something more powerful than anything she had ever felt and fought to control herself. "I believe she was a Crossbearer, one of those dissidents who cause so much trouble."

"They call themselves Christians," Jim said.

"But what was she talking about?"

"She saw the beauty beyond this earth."

"I don't understand," Starr replied disconcertedly.

"Nor do I, Starr Omega. Nor do I," Jim said as he returned to his easel. "But I know it's there."

"How do you know?"

"I've seen it in faces — and in lives."

Jim began to work on the painting and Starr remained in her chair, utterly perplexed. One thing she knew. She had to see the world outside the Dome! She had no idea how long she sat there, but before she told Jim goodbye, she determined how she would accomplish this thing that was now the driving force in her life.

Bernard Alpha was stretched out on the cool, green grass in the shade of a willow listening to the sound of the children at play. He opened his eyes and saw how the sunlight touched the leaves above him with gold. The sound of water trickling over stones in the streambed came to him like soft music. He sighed deep in his throat, closed his eyes and was drifting, into a deep, dreamless. . . .

"Starr Omega here to see you, sir."

The sound hit him like an electric shock. He quickly replaced the black and white print, closed the desk drawer and locked it. "Yes, yes. Send her in."

Bernard watched Starr enter and take a chair. There was an intensity in her face and actions that he had never seen before. "I can't imagine what would bring you to this dreary place on your afternoon off, Starr. It must be urgent."

"Not exactly," Starr said, leaning forward in her chair, "but it could be exciting."

"If it excites you, Starr, it must be something truly cataclysmic," Bernard said. "Don't keep me in suspense."

"Well, I think I've come up with something that will kill two birds with one stone, to use an ancient phrase. It has to do with the Chr—," Starr said, stumbling over the word. "Crossbearers."

Bernard had never seen Starr so animated. He noticed the flushed appearance of her face, then asked, "What about them?"

"As you know, I haven't decided on a subject for my dissertation yet," Starr replied. "I believe I can complete it on the 'Crossbearers' and render a service to the City at the same time."

"I'm intrigued, Starr," Bernard said. "Do go on."

Starr knew she must be cautious and choose her words precisely from this point on. "We all know this spurious philosophy is beginning to cause some problems in the Rim. For example, these people continue to congregate when they know this is expressly forbidden. Freud only knows what goes on at those meetings of theirs. There are rumors that the insidious effects of this — this blight have reached the Fringe and. . . ."

"Starr," Bernard interrupted her, "I'm quite familiar with the goings on in the City. Will you get to the point?"

"Yes sir. It seems that the stronghold of this — this. . . ."

"Christianity, I believe, is the word you're looking for, Starr," Bernard interjected.

"Yes, thank you." Starr was grateful he had used the word first. "The stronghold of this Christianity appears to be in the Fields. Word has it that even some of the Border Guards have fallen victim to its outrageous promises."

"That's quite correct," Bernard said. "The Border Guards are a different breed. Theirs is a profession that's been handed down from father to son for generations. Every precaution is taken to ensure their absolute loyalty to the City. Even so, some of them have succumbed to Christianity."

Starr had her opening. "This is precisely why I'm so excited about my project. If this disease can infect even the strongest of us, something must be done to eliminate it. In order to devise a cure, we must first discover what causes it. What I propose to do is to travel into the Fields, gather the essential facts and use them to implement a final solution to the problem."

Bernard saw the logic of Starr's proposal. He also saw the benefits for him if he were to sanction it and it proved successful. "How much do you know about the Fields?"

"Almost nothing," Starr lied.

"There are dangers, environmental and others, such as criminals that roam at will. The Border Guards can assist you in remediating them of course, but you will still be in some peril."

"It will be worth the risk if I can be of service to the City," Starr lied again.

"Have you cleared this with your advisor at the university?"

"Not yet," Starr said, trying to control her elation that things were going even better than she had expected. "I know that the White Tower must give approval for all trips outside the Dome. You're the proper authority to accomplish this."

"I can see you know your politics, Starr," Bernard said. "If I were to champion your cause in the White Tower, and persuade them to sanction it, that would override any objections you might encounter elsewhere. Most efficient."

"Thank you, sir."

"I'll take your proposal under advisement," Bernard said. "You'll hear from me soon."

"I appreciate your help, sir," Starr said, rising from her chair. As she left the office she saw that marvelous light streaming through the trees and felt its warmth on her like the caress of her mother. Her spirit soared within her as she walked through the fragrant, green Fields toward another life.

The florescent light gleamed on Richard Xi's close-cropped white hair as he paced behind his desk. He reminded Bernard Alpha of the poet's description of another Richard, "clean favored and imperially slim."

"How do you know we can trust this Starr Omega?"

Bernard had been in the White Tower for thirty minutes and he couldn't wait to get out. After Starr had left his office, he arranged an appointment that evening with his friend, Richard Xi, in his office in the White Tower. He had known Richard for years (two of which were as Loving Friends), and knew he could persuade him to obtain the necessary approval. "You've read her file," he said. "On the same day last week she had two assignments who were Christians. She carried them out admirably."

Richard's eyes were dark as a ferret's as he

looked at the monitor on his desk. “I also saw your notation of that date expressing some doubt as to her steadfastness in this area,” he said.

“That was before the field assignment with the woman named Martha Epsilon,” Bernard Alpha said with conviction. “The way she handled that erased all doubts as to her steadfastness, as you put it. It was her ‘Rite of Passage’ as a Reliever.”

“I can certainly see the merit of her project,” Richard Xi responded. “This Christianity business needs tending to and you’ve convinced me of her resoluteness. I’ll see that the proper clearances are taken care of. I don’t foresee any problems.”

“You’re a real friend, Richard,” Bernard Alpha said as he rose to leave. “You won’t be disappointed.”

“I’m sure I won’t, Bernard. You take care, ‘Old Friend.’ ”

As soon as the door closed, Richard Xi retrieved Bernard Alpha’s personnel file from Central Records and typed the essence of their meeting in the form of a legal document. When he finished he looked at the screen and satisfied himself as to its accuracy. As an appendage he typed:

Respondent has agreed to accept full responsibility for the actions of his employee,

Starr Omega. Should they go contrary to the best interests of the City — he is to be terminated immediately.

Sweat and horse manure and sore muscles. This was what Starr remembered most as she readied herself for inspection on this final morning of her four week orientation at the Camp. It wasn't really a camp at all, but the Border Guards had called it that for as long as any of them could remember.

The Camp was a walled off area inside the Dome just north of the Main or Western Gate (the name the guards used). It contained a one-eighth mile jogging track, whose infield was used as a parade ground and physical training area; a gym, complete with exercise equipment, steam baths and four racquetball courts; an obstacle course; and a firing range. There were also stables for the horses and barracks for the men.

The guards lived with their families in an area very similar to base housing, located adjacent to the Camp. The barracks were used by new guards during their six months of Basic Training and there was one transition barracks for the Regulars who were going on and coming off duty. This arrangement made for convenience as well as a sense of unity for the guards and their families. It also gave the

City control over this strategic and potentially dangerous body of men called Border Guards. Every guard on duty in the Fields knew his family remained under the watchful eyes of the Peacemakers.

"You're certainly a different girl from the one who showed up here four weeks ago. There's a little color in those cheeks and I do believe you've learned to stand up straight," Michael Kappa said as he walked around Starr, hands clasped behind his back. "Yes sir, you look downright healthy."

Starr did feel healthy for the first time in years, in spite of the aches and pains she had endured. Michael had been her personal instructor for the abbreviated training course and she had thought him a sadist for the first two weeks. Blonde and crew-cut, with eyes like gun metal, he reminded her of a Marine recruiting poster she had run across in her research. He had been relentless with her on the jogging track, the obstacle course and in the gym. Outside the Dome, she had ridden the horse until she could barely climb down from the saddle. Then there were the three days of wilderness survival — she didn't even want to think about that!

After a complete check of her horse and equipment, Michael returned to stand in front of her. "I guess you'll do, Starr Omega.

You've come a long way in a short time," he said with a smile.

He actually smiled, Starr thought.

"I still think you should wear the standard Guard issue," he said, looking at her Levi's. "Some things never change do they? I expect they're durable enough though. At ease."

Starr took a deep breath and relaxed. Then she smiled back at Michael. "Have I been too much trouble?"

"You whined a little at first, but then so do most of the men. Actually it's been kind of interesting."

"Do you think I'll make it out there, Sergeant?"

"It's Michael now, Starr. Training's over," he replied. "I don't see why not. You've got plenty of rations, maps in your computer, a fine horse and you're in pretty good shape."

"I guess I'll be all right," Starr said, taking another deep breath.

Michael noticed the hint of a frown crossing her face. "Don't worry about the criminal element, Starr. Most of the Primitives are gentle, hard-working people. You may even grow to like some of them. Just don't attract attention to yourself before you get settled in one of the villages. You'll be fine then."

"I appreciate all you've done for me, Michael."

"Forget it. You've put me in the history books. No woman's ever graduated from here before."

"Do I detect a note of Chauvinism in this enlightened age of ours, Michael Kappa?" Starr said with feigned chagrin.

Michael smiled broadly. "Just a tad, maybe. By the way, your clearance came through this morning."

Starr was elated. "Oh Michael, that's wonderful! When do I leave?"

"Tonight."

Starr was visibly shaken. Now it was real. Tonight she would enter what most people thought of as a land of darkness and terror at the end of the world.

Michael stepped close to her and put his hands on her shoulders. "You're gonna do just fine out there, Starr. You've got what it takes," he said and put his arms around her.

Starr hugged him back and felt a strength and reassurance in his concern for her. She wondered what it must have been like to have grown up in a family and to have had a brother.

Part Two:

THE FIELDS

Chapter Six

Color And Light

Holding the reins, Starr stood next to the palomino and watched the massive door of the Dome slide downward on its tracks and slam heavily into place. Putting her left foot into the stirrup, she swung into the saddle. She took a last look at the huge glowing bulk of the Dome that sloped up and away from her until it disappeared into the night sky. Then she turned her face toward the west and rode away toward the unknown.

The sound of the River came to Starr — a soft lapping of the waves on the mud bank. She had never seen a river except for a few scenes in the central files that had escaped the sensors, and she wished it was day so she could see this one. But she smelled it — smelled the cool, damp breath of it as she approached and saw the white mist that reached out from it all about her, as high as the horse's fetlocks.

"Who goes there?" a voice called out of the night.

"Starr Omega," she replied simply.

"We know of your journey." A man seemed to materialize twenty feet in front of her. He had on a rough brown jacket and pants of the same material tucked inside high-topped leather boots. A wide leather belt and leather straps that crossed his chest were laden with shells for the shotgun that hung on his right shoulder. "Follow me."

Starr dismounted and led the horse behind the man. When they reached the River, she saw a metal barge with a gangway leading up to it from the bank. The man motioned to her and she led the horse up through a gate in the railing that encircled the barge. She saw the man walk to a pole that stood in a concrete slab near the gangway and pull a lever up. There was a sudden jerk and a shuddering; then the sound of pulleys creaking and the barge began moving slowly toward the opposite shore of the River.

The waves slapped against the hull and there was a shriek of metal as the cable scraped against it. The horse neighed and shook his head, but Starr rubbed his neck and spoke softly to him and he settled down. The mist had thickened on the River and time and distance seemed to lose meaning. The shoreline appeared abruptly and the barge slid to a stop on the muddy bottom.

The gangway, which was on the opposite end from where she had boarded, lowered automatically and Starr led the horse carefully onto the bank. Then, as the gangway raised, the barge slowly backed into the River and was lost in the mist. Starr led the horse up a slope, feeling the grass brush against her boots. She paused at the top and bent down to feel the cool grass with her hands as the night wind blew unfamiliar fragrances to her.

The horse began grazing and Starr stood up and looked back across the River toward the only life she had ever known, as slaves must have done when they were freed from the plantations. Leading the horse down the slope, she felt an ill-defined fear in her breast, but there was also something deeper and stronger there; a kind of hunger, a longing — for what, she didn't know.

At the top of a second slope, a voice came from a speaker inside one of two black columns in front of her. "When this message is completed, you will have ten seconds to pass between the columns. Do not hesitate! A prosperous journey and safe return to the City, Starr Omega." A harsh buzzing sounded the end of the message.

Starr hurried through the laser fence and mounted the horse. As she rode, she reached behind her into the saddlebag for the computer. She flipped the cover open and the familiar glow seemed to comfort her. After calling up the section map for the area she was riding through, she decided to follow the creek south of her that ran almost due west.

Riding through the night, she listened to the sound of the creek and the wind blowing through the trees on its bank. There were frogs croaking, the cries of night birds and animals rustling in the underbrush. It was a continual source of fascination for Starr, and she had no trouble staying awake.

Around 2:00 in the morning she decided to give the horse a rest and stopped beneath a willow on the bank of the creek. She gathered some limbs for a small fire and lay with her back against the tree, watching the flames flicker in the breeze and the bright red glow of the coals.

Sleep had just begun to take her when the thunderous roar of a shotgun blast shook the night. Starr was fully awake and standing with her back against the tree when twenty yards in front of her a tongue of flame appeared and the shotgun roared again. She ran to the horse, who was rearing and snorting, and grabbed the reins.

A tall man dressed like the one she had seen at the River appeared at the edge of the circle of light. A large dark horse stood behind him, apparently undisturbed by the sound of the shots. The man had sandy hair and was almost bald on top, but the back and sides grew down to his shoulders. His beard was reddish blonde and was almost as wild as the look in his eyes. He held the shotgun in his right hand, pointing it at the earth.

"Starr Omega, I presume," he said as he smiled with his mouth and his eyes.

Starr was dumbstruck. She held onto the neck of the horse to keep from falling down.

"I heard that somewhere a long time ago — or maybe I read it. Anyway, I liked the sound of it. *I presume.* It has a certain ring to it, doesn't it?"

Starr was coming back from sheer terror to ordinary fear. "Why did you fire that awful gun?"

"This?" the man said, holding the shotgun out to her like an offering. "Why, I like the sound of it too. Tends to liven up a dull evening on the prairie, don't you think?"

Starr could think of nothing else to say.

The man led his horse over to Starr's and tied him to the sapling. Then he stepped to the fire and dropped his rucksack on the ground along with two dead rabbits he had been holding in his left hand. He built up the fire with some limbs she had gathered earlier and picked up the two rabbits. Pointing to them, he said, "Saw them in the edge of your firelight. Makes their little eyes glow red just like the coals in the fire."

Starr seemed unable to let go of the horse. "Who are you?" she asked. "What do you want?"

"Name's Will Sigma," he replied, as he pulled a long thick-bladed knife from a sheath at his belt.

Starr gasped.

The man laughed a deep rumbling laugh that made his eyes smile again. "And what I want is to eat these two lovelies," he said, walking the few feet to the River.

Starr felt the fear leave her. *He's had enough chances to harm me if that's what he intended,* she thought. She walked over to the creek bank where Will Sigma was kneeling and cleaning the rabbits.

He looked over his shoulder at her. "You're welcome to join me," he said with a deep voice.

She looked at his bloody hands and made a face.

Will laughed again. "Not with the cleaning of 'em, girl. With the eating of 'em."

Starr sat on the ground, watching him skewer the rabbits on a limb he had cut with one stroke of the big knife, then roast them over the fire. He had added salt and pepper and she thought she had never tasted anything so delicious.

When the meal was finished, Will threw the scraps into the River and made coffee in a small drip pot he took from the rucksack. Starr held the cup in both hands and smelled the rich aroma as the night sounds began to die out around them.

"What do you do out here?" she asked Will.

"Have a whale of a good time mostly," he said as he sipped the steaming coffee.

Starr smiled and thought that she had never seen anyone who enjoyed the simple fact of being alive as this man did. His presence seemed to brighten the night the way the fire did. *Does this wild open country do that to people?*

"I took this season's Quota Charts out to the villages in my sector. I'm on my way back to the City," he said.

"Will they let you stay there for awhile?" Starr asked.

"Not if I can help it," he said. "I can't breathe in that place and the food." He made a face and spat on the ground.

Starr looked at the shotgun and the knife. "What are these Primitives like?"

Will laughed again. "Not dangerous if that's what you mean. Most of 'em anyway. Some of 'em plunder their way through life though, like they would anywhere they were put down. There's a certain type of man who'd rather die than work for a living."

"I've heard they're mostly criminals and people who tried to overthrow the City government. The misfits of society."

"*Misfits.* I guess that pretty well describes them. They certainly don't fit in that place you come from." Will stretched his long legs and leaned back on his elbows.

"I shouldn't have any problem with them then?" Starr asked.

Will looked at her and his face was serious. "Some people change after they're out there for awhile."

"What do you mean?"

"It's hard to say exactly. Some of these primitives have insane ideas. They say you should love your enemies. Can you imagine that? They treat everybody pretty much the same. No big shots. I know you can't imagine that!"

Starr took the computer out of her saddlebag and began taking notes. "What else?"

"Well, once or twice a week they all get together and sing and somebody will talk to the others for awhile. And — other things that are hard to explain."

"Try to tell me about these — other things," Starr said, looking up from her computer.

"You'll see that for yourself," he said. "Everything is done in families. They raise their own children, live together, work together, eat together — everything."

"Go on."

"Can't do it, little girl. Got to get my report in. You ought to know how stuffy they can be if you're late with anything. Enjoyed your hospitality though."

Will washed the coffee pot and cups in the creek and packed them in his rucksack. Then he climbed on his horse that was tied next to Starr's. "You'll do all right out here," he said.

"Thanks for the meal," Starr said. "It was wonderful. You're a nice man."

"There's some who might disagree with you," Will said. "Don't get taken in by them Primitives, Starr Omega. *Love your enemies.* Can you imagine that?"

After Will Sigma was gone, Starr watched the embers dying in the fire and felt a strange sort of emptiness. The breeze was softening with the approach of morning and she decided to be on her way. She saddled the horse and rode away from the approaching sun.

Starr came to the top of a long rise and halted the horse. The sun was rising behind her and the first streaks of light broke over the horizon and swept down across the long gradual slope of the valley spread out before her. The tall wet grass seemed to ignite with a green-gold flame as the sunlight struck it, and, as far as she could see, it billowed and glistened in the morning breeze like a sea of fire and ice.

The colors were so vivid and brilliant they hurt Starr's eyes and made her dizzy. She closed her eyes and let the horse take her down the slope, leaving a wake behind him in the dew-wet grass. The words of Martha Epsilon came back to her, "the Fields. The beautiful green Fields where I was born." Starr opened her eyes and was dazzled by the waving glistening greenness around her. *This is what she meant,* Starr thought. *Now I understand what she meant.*

Starr thought of an old movie she had run across in the archives. It was about a little girl who was caught up in a tornado in this very part of the country and taken to a magical kingdom. The first part was in black and white, but when she reached the kingdom, everything was alive with glorious color. Starr felt she had lived in a black and white world until the sun rose on this day.

The smooth gait of the horse took them through the endless rolling fields of green with occasional streams and stands of huge trees. The sky was a cobalt ceiling curving from green horizon to green horizon. Puffy white clouds followed their shadows across the land like balloons being towed by invisible strings.

At the top of another rise, Starr saw three men plowing the green land into straight dark furrows. As she rode down toward them, one man left his horse and plow and walked toward the shade of a tree that towered alone in the midst of the plowed field. Starr followed him on the horse.

As Starr approached the man, she saw that he wore heavy boots, green work pants and a white cotton shirt. "Good day to you," she said formally.

The man turned quickly and looked up at her. "Oh, hello there," he said. "I'm afraid I didn't notice you. I sometimes have a tendency to get lost in the work."

"I'm sorry. I didn't mean to startle you."

"Think nothing of it. It's good to see a new face out here." He dipped water from a wooden bucket with what looked like some sort of a wooden cup with a long hollow handle. "Have some. It's very good," he said as he offered it to her.

Starr stepped down from the saddle and walked over to the shade of the tree where the man stood. She took the dipper and drank the sweet cool water. "Thank you," she said.

The man dipped again and walked over to the foot of the tall tree. He sat on the leafy ground between two roots that extended from the base of the tree like the arms of a chair. "Rest with me," he said. "You look like you've ridden a long way."

Starr sat in the cool shade under the tree and looked out across the grassy fields where the men were turning them from green to a rich dark earth color with their horses and plows. The wind sent gentle ripples across the surface of the grass and set the leaves rustling in the crown of the tree.

The man drank slowly with great sighs of pleasure. He was very dark with a small thin nose and brown liquid eyes. "My name is Obadiah," he said. "Call me Obie."

"All right, Obie." Starr said smiling at the small man. "I'm Starr Omega."

He extended his hand and she took it awkwardly as he gave it a gentle shake.

"Could you tell me what that is you're drinking from?" Starr asked, pointing at the dipper in his hand. "I don't believe I've ever seen anything like it."

He held it up in his thin hard-looking hands. "It's a gourd. We cut 'em, dry 'em and make dippers out of 'em."

Starr took it and ran her hands over its smooth, hard lightness. "It's lovely," she said.

"Like a piece of sculpture."

"It's yours."

Starr was shocked. No one had ever given her anything without expecting something in return. "I couldn't," she said, handing the gourd back to him.

There was a hurt look on Obie's face. "Keep it," he said. "As a token of friendship."

"Friendship," she said, then added, "I'm doing research for a paper I'm writing. Could you tell me where the closest village is?"

"That's easy," Obie said as he stood up. "The one I live in." He pointed due west.

"Is it very far?" Starr asked.

"You stop for the night and you'll reach it before dark tomorrow," he replied. "You could travel back with us but we stay out three or four days when we work the land."

Starr thought of asking him about the "religion" she had heard of among the primitives, but didn't want to put anyone on the defensive. Better to just wait and see for herself. "Thank you for the gift, Obie," she said.

"You're very welcome," he replied and walked toward his horse and plow in the green, ever darkening fields.

Starr rode for the rest of the day through the same kind of country. She never tired of the wonderful wild beauty of the land. She saw deer once or twice in the wooded areas near the streams and watched a hawk circling between the earth and the clouds, riding the thermals and never moving his wings.

She made camp at the crest of a hill, deciding not to build a fire. She ate some of the awful-tasting dried food that she had brought with her from the City, while the horse grazed on the abundant grass. As the night wind whispered across the hills, she lay back to rest and was paralyzed by the unexpected, majestic beauty of the night skies. She realized she had never looked up in the excitement of the first night. The great black dome of heaven was covered with thousands and thousands of bright twinkling stars and the shining of the planets. The clarity was such that she felt she could reach up and pluck one from the sky. She could have lain there all night and looked at them, but there was work to be done.

Starr opened the computer and wrote of the first day's activities and observations in the formal and stilted language of academia. Then she took a yellow pencil and a small tablet with a thick brown cover from her inside jacket pocket and began to write.

> The countryside is beautiful beyond belief. The air and water are pure as far as

I can tell (Why the Dome?).

Met a Border Guard the first night. An exciting, entertaining character. Shot and cooked rabbits that tasted better than any restaurant food. Loves being in the Fields. Would never return to the City under any circumstances.

On the first day I saw men plowing the fields with horses. Met one named Obadiah (Obie). Very kind and gentle. Gave me a gourd as a present. (Is this a common practice?) Also gave me water. Giving may be a prevalent custom.

Should reach Obie's village tomorrow morning.

People fascinating. Land pristine and lovely.

Danger — Possibility of insidious effects from this (Philosophy? Religion?) practiced by the Primitives. Horse has performed admirably.

Chapter Seven

Storm

Awakening at first light, Starr realized she had gone to sleep while writing and slept the night through without taking a pill. *What an incredible sensation!* she thought. *I'm beginning to like this place in spite of myself.* The dry breakfast packet she ate put a damper on the morning, but she soon forgot it as the glowing rim of the eastern sky gave way to the dazzling blue and green beauty of the day.

Prairie flowers were as high as the flanks of the horse and flowed in waves of amber and blue across the gentle roll of the land. Starr rode all morning as if in a dream, stopping only to water the horse at an occasional tree-bordered stream that wandered through the Fields. As the day wore on, clouds began building on the western horizon and she watched them climbing slowly toward the sun.

Toward evening, Starr came to a rocky outcrop that dropped off fifty feet to a small valley below. Two hundred yards to the north, a steep path wound down to the valley floor. As she reached the last steep incline at the bottom of the path, a long black snake slithered from under a rock ledge startling the horse. He reared, throwing her to the hard-packed ground, and bolted away down the valley.

Starr lay on the ground in a daze. Bright flashes of color whirled in her head like a windless cyclone, and the colors were the pain that she felt as she lay on the path in the fading light.

Her head was throbbing and her left hip was stiff and painful as she awoke. She saw first the warm steady shine of the evening star and watched it wink out as the clouds moved toward her. Then the wind came sighing through the tops of the trees and the first heavy raindrops hit, shocking her with their stinging coldness.

Life in the climate-controlled Dome had not prepared Starr for the capricious turns of nature. She struggled to her feet and looked about. To her left was a dark arca in the wall of the canyon. She walked stiffly toward it with her hip aching and pain radiating from the small lump at the back of her head.

The ceiling of the cave was high enough that Starr could not touch it with her hands. It was ten feet wide and six feet deep and there was a stack of firewood against the left wall. She placed some of the wood in a shallow pit, toward the rear of the cave, that was blackened from the fires of those who had been here before her. There were twelve matches left in the waterproof cylinder she carried in her coat pocket. She laid it on the stone floor of the cave and cut a small pile of shavings with her belt knife. They ignited with the first match and she added more until she had a small flame going. Adding broken twigs and larger pieces of wood, she soon had a respectable fire.

The rain sounded like a waterfall out in the darkness and the inconstant wind blew a cold spray through the mouth of the cave. Starr sat on the damp floor leaning back against the rough stone and watched flickering shadows play across smoke-blackened walls.

No! The City had not prepared her for this: this thundering darkness, this howling wind that seemed to have a life of its own as it snapped at the flames of her small fire like some enraged and formless beast. She huddled close to the fire's warmth. This was nothing like the open fields and sunlight and flowers. This was dampness and cold and decay. *Like a tomb,* she thought.

Most of all, Starr was not prepared for this singular and unrelenting aloneness. She had lived all her life in the company of others; the privacy of her apartment had become an obsession with her. Now she would give anything for the sound of a human voice. She felt she was alone, hurtling through space on a planet devoid of life, except for her own.

Then it hit her. Hit her with a sudden, brutal clarity: *This is death! This is what death is like! This total and unbearable aloneness. This suffocating darkness and the formless beast living in it, living on the pain of death. And there is no end to death — there is only eternal terror and pain and darkness.*

She suddenly knew with absolute certainty there was no endless recycling of the soul into body after reincarnated body. There was an eternity out there, but this life was the only one she would have to face it with. There was only one Starr Omega and some part of her would be alive somewhere forever after this life was over.

She thought of Philemon and of Martha Epsilon. How could they face this terrible empty darkness knowing there would be no other life in some other body? Where does such calmness and peace come from? Only with the most potent drugs had Starr seen anything approaching this and then only temporarily.

Lost in thought, it was some time before

Starr realized the sound in the storm was the neighing of a horse. She was about to rise when a span of darkness deeper than the night loomed outside the mouth of the cave. A man stood at the edge of the firelight, then stepped out of the rain, his bulk filling the mouth of the cave. A heart-stopping fear washed over her when she saw him in the light.

His shaggy dark head nearly touched the ceiling of the cave and the heavy beard that grew almost to his eyes was matted and filthy. He wore a coarse black robe that hung below his knees, touching the heavy boots that were caked with mud. Water was dripping from his clothes to the floor of the cave and the smell that came from him was like the last stages of some terrible rotting disease.

But the eyes were the worst! They were lifeless and cold and appeared to be covered with an opaque film, like a reptile. He glanced at Starr as though she were no more than an insect and stepped to the fire to warm himself. She scuffled backwards and sat huddled against the far wall.

The man held his hands over the fire and Starr noticed the skin under the dark hair was stained and grimy, the long nails caked with dirt. He hawked up a great gob of yellow phlegm and spat it hissing into the fire. Squatting down, he pulled a bone with some scraps of meat and gristle clinging to it from inside his coat and began gnawing on it. So great was her fear of this man, Starr wished for the aloneness of the storm and the night. She gathered the remnants of her courage and said in a rasping voice she hardly recognized as her own, "What do you want?"

The man continued to gnaw on the bone.

"What do you want?" she asked again more strongly.

The man never looked up from his eating. "You make a lot of noise for a dead woman."

His words were messengers from the nether world that struck Starr with disbelief, sending a numbness crawling over her body like a swarm of hideous spiders. She felt herself being drawn into a dark pit and fought against it. If she lost consciousness, she knew that she would never awaken.

Starr decided to try one more time. "Please don't hurt me. I'll give you everything I have."

The man threw the bone into the darkness and looked at Starr. In the flat, dull saurian eyes she saw her life wink out like the evening star before the storm.

"It's already mine," he said.

The man stood and pulled a knife from under his cloak. It was as long as his forearm and glinted coldly in the light of the fire. Starr saw scenes of her life flashing past. She saw Martha Epsilon as her head moved slowly to the side when she fell asleep. Then the man was standing above her and the knife in the grimy hand with its black nails was moving toward her.

"Oh, God — help me!" Starr screamed, though she didn't realize what she was saying.

The actions that occurred next were almost simultaneous; only later was Starr able to piece them together. There was a blur of motion and a dull cracking sound as the bones in the hand that held the knife were shattered. The knife clattered to the stone floor and the man in the black cloak bellowed in pain and rage.

Another man stood in the cave now, facing the man with the crushed hand. He was as tall, but had none of the bulk of the man he faced. A staff of polished oak five feet long and the size of a woman's wrist was balanced lightly in his hands. A leopard moves toward his prey as he moved when he circled to his right, coming between Starr and the man with the crushed hand. Starr saw his broad shoulders in the gray cloak and the sheen of his long black hair as she scrambled to her feet and stood against the wall behind him.

The man in the black cloak looked at his long knife lying by the fire, but the man who moved like a great cat pointed his oak staff at the bearded face and moved it slowly back and forth.

Then the man with the crushed hand held it with his good one and backed out of the cave. "The hand will heal," he said and disappeared like a dark vapor into the rain and the howling wind.

The tall man in the gray cloak turned and faced Starr. She looked into his dark blue eyes and saw him smile. "My name is David," he said. "You don't have to be afraid now."

And Starr knew that she would never have to be afraid with this man as relief washed over her like the rain. She stepped to him and the fear broke loose in deep gasping sobs that rose from her chest. He put his arms around her, and blindly, without thought, she clung to him until the fear drained from her. She stayed there in the comfort of his arms for a long time, and she wept like a small child.

After she was taken from her parents, Starr discarded her dependency as she had the toys of her abbreviated childhood. She came to develop a certain pride in her self-sufficiency and kept a carefully measured distance from the lives of others. She was not a people-hater, but determined during her first years in the Caretaker system that nothing was worth the pain she had felt with the loss of her mother and father.

Standing in the arms of David, Starr sensed a battle beginning behind the wall she had built around herself. Years of self-discipline told her to break away, to stand apart, that no good could come of this. The warmth of his arms around her and the feeling deep within her breast told her something utterly different.

The years of her life seemed to flow backward as Starr searched them for some spark of meaning, some moment of true happiness. She felt she had been all her life a gourd, dry and useless, never knowing it until now — now as she was filled with a clear sweet water. She longed to let go, to release the aching burden of the years, to give that part of herself that was dying. She could not! The wall was built strongly and well during all the long nights.

David felt Starr's body tensing in his embrace, felt her hands pressing against his chest, but he didn't feel the dark anguish of her soul as she pushed him away.

While Starr sat watching the fire, David disappeared into the night and returned five minutes later with a canvas bag over his shoulder. He took two blankets from the bag and laid them by the wall where Starr sat.

Then he took a skillet and several packets wrapped in oilcloth and placed them next to the fire. In a short time he had some red meat simmering in a gravy on the fire.

The smell of the food was intoxicating to Starr. She lay back on the blankets, luxuriating in the warmth and safety she felt with this man. David handed her a tin plate of meat and brown bread along with a metal knife and a fork carved from wood. Then he served his plate and sat with her on the blankets.

The food was even better than Will's. As she ate, Starr found herself staring at a white scar that cut through the tan skin along David's left cheek. It extended from just below the eye directly down the prominent cheekbone to the corner of the mouth. It reminded her of the Prussian dueling scars she had seen in the old texts while doing research on Adolf Hitler.

"It's a gift from your uninvited house guest," he said, looking up. "I'll always remember him fondly for it. His name is Abbadon and there aren't many around like him, thank God."

"I'm sorry. I didn't mean to stare," Starr said. "Why did you allow him to leave?"

"He was helpless."

This made no sense to Starr. She looked at

Abbadon's knife lying next to the fire and decided to change the subject. "I had another meal very much like this last night," she said. (Was it only last night?) "It was prepared by a man named Will."

"I know him. He's a good man," David said as he looked up from his food.

Starr looked into the steady blue eyes that seemed to reveal the man. How can he say this of one of his captors? she thought. "Do people out here eat like this all the time?" she asked.

"On the move we do. It's much better in the villages," he said. "What do you eat?"

Starr handed him one of the food packets. "This was prepared for my journey. It has all the nutrition necessary to sustain one," Starr said formally.

David opened the packet and tasted the green pasty substance inside. "It also has the smell and taste necessary to make one lose his supper," he said, handing the packet back.

"Food is required for the maintenance of the body," Starr replied defensively. "We derive pleasure in ways you never dreamed of."

David finished his meal, set the plate aside and leaned back on the blankets. "Such as?" he asked.

Starr took a small plastic box from her

inside coat pocket, slid the top open and offered it to David. "Try one of these and you'll find out," she said.

David looked at the thin white wafers inside the box. "What are they?" he asked.

"They make you feel good."

"I already feel good."

Starr pushed the box at him insistently. "Try one. What are you afraid of?"

"Nothing I can think of right now," he replied.

"Look. I'll show you," Starr said. "It'll heighten your pleasure." She took one of the wafers and placed it on her tongue.

David watched her with curiosity.

Starr felt the drug rush through her system and saw David getting smaller and smaller as she lifted above him. She seemed to be drifting outward into the darkness and, by a supreme effort, willed herself back beside him on the blankets. Starr felt there was a great gulf between her and David and she desperately wanted to cross it, desperately wanted to be where there was contentment and peace, but she didn't know how. She felt her will go and began quietly weeping.

David put his arm around her and laid her head against his shoulder. "There, there. It's all right," he said. "A person can stand just so much pleasure."

Starr awakened in the dead of night. David lay next to her on the blanket, breathing deeply in his sleep. She marvelled that he had made no advances toward her, but expected it soon. In fact, as she looked at his lean tanned face with its dueling scar, she was excited at the prospect. *He's merely a Primitive,* she thought. *How can I possibly have these feelings about him?*

The storm was over and she heard the ticking sound of the rain dripping in the trees. In the distance, frogs were croaking and the cicadas were droning their nocturnal symphony. The air was fresh and cool and sweet. Starr felt that the coming of Abbadon was something out of a nightmare, something that could never truly happen in this clean, wonderful land.

Starr smiled in the dark and stretched herself in the warm blankets. "Are you all right now?" she heard David say.

"I'm just fine," she answered.

David stood up and put more wood on the fire. He smiled down at Starr as he stoked it, and the coals sparked and glowed brightly. "I'm glad," he said.

He lay back down and Starr turned slowly toward him, determined to put him in his place when he reached for her. As she waited, she saw him turn away from her and he was soon asleep.

What's the matter with him? she thought. *What's the matter with me?*

The morning came to Starr with the sound of birdsong and the rich smell of coffee brewing. She sat up and yawned, surprised that she had rested so well with only a blanket between her and the stone floor of the cave. The sun was sparkling in the wet trees and on the long grass down the valley.

David handed her a tin cup of coffee and she held it in both hands, sipping it slowly. "How do you manage to get coffee out here?" she asked. "It's hard enough to come by in the City."

"We have a few contacts," he replied, looking out of the cave at the bright morning.

Starr looked at David as he sat wrapped in the plain gray cloak he had dried by the fire. There was something regal in his relaxed, confident posture. "You haven't asked me why I'm out here in the Fields," she said. "Aren't you interested?"

"You'll tell me if you want me to know," he said, looking at her with the rumor of a smile on his face.

"I'm doing reasearch for my doctorate degree," she said. "Are you familiar with that?"

"Somewhat."

It struck Starr that this was the same reply she had made to Sheila Phi's question about the work of Gertrude Stein. "No matter. A man named Obie told me there was a village near here. That's where I'm going. Are you familiar with it?"

"It's called Haven," he replied. "That's where I live. We'll be there by noon."

David had set their dishes outside in the rain after supper and he now dried them and put everything away in his canvas bag. "I'll go get my horse," he said. "By the way, he had some company during the night. A palomino."

Starr's face brightened as she followed him outside. Twenty yards down from the cave, there was a shallow overhang in the cliff with two oaks growing at the edge, their limbs spreading over it forming a natural shelter. Starr's palomino stood next to the chestnut stallion belonging to David.

"He came in during the night," David said. "Doesn't seem to be in any hurry to leave."

Starr ran to the horse, spoke to him and patted his neck. "I didn't realize how attached to him I'd become," she said. "It's the first time I've been around an animal."

As they rode through the open countryside, Starr became intrigued with what kind of place the village would be. "Will there be a place for me to stay?" she asked David. "I have Credits."

"What would we use them for?"

"I hadn't considered that, I guess."

"Don't worry. We'll find a place for you." David reassured her. "You'll have to work though."

"What kind of work?"

"Whatever you can do," he said. "Everyone does some kind of work in Haven."

As they rode, the day warmed and David tied his cloak behind his saddle. Starr took off her coat and put it in her saddlebag. The sky was a rinsed blue after the rain and perfectly clear. A wind was freshening from the west and the long grass was billowing and glistening in the morning sun.

At the top of a rise, Starr noticed what looked like a gigantic pile of rubble several miles to the north across the open fields. "What's that?" she asked, pointing to the north.

"What's left of a city. You do know they destroyed all the cities inside the laser wall, don't you?"

Starr knew, but she had never seen any pictures — could never have imagined such complete destruction. "I guess I'd forgotten," she said, looking away from the ruined city.

"The government of the City didn't want

us to have any of the modern conveniences: electricity, running water, no machinery of any kind. They even tried to destroy all the books. Thought it would people their 'Gulag System' with sub-humans, beasts of burden incapable of rational thought. I find it has proven a blessing for us."

"In what way?" Starr asked.

"It's difficult to explain and it hasn't been in every way — medical care for instance is pitifully inadequate. But there are things more important than medical care."

Starr pondered on this man David and what she might expect in his village. He was strong enough to defeat a man like Abbadon, who could easily handle anyone in the Peacemakers. Yet he had let this man live for no good reason she could think of. He was gentle and considerate with her and didn't try to press his advantage during the night. She tried to imagine Sammy Chi in the same situation — then blocked it from her mind.

David never considered using drugs and he seemed to have limitless courage without them. Starr knew he had no opportunity for a formal education, yet he appeared in some ways more educated than she. She decided to look further into this enigma. As she watched his relaxed, controlled movement astride the stallion, she thought of something Will had said of some of the Primitives. She couldn't speak for the rest, but David would most assuredly be a "misfit" in the City.

They had come to a small river and crossed it at a shallow place with a rocky bottom. Far to the north, she could see where it joined a much larger river. North of the crossing, the land began to rise and became a high bluff with its apex where the rivers joined. The entire area, bordered on the east and north by the two rivers, was heavily forested with towering oaks, smaller trees and various kinds of shrubs.

They followed a path through the trees that ran close to the rising bluff that looked down on the smaller river. Gradually the forest began to open up and the underbrush and smaller trees had been cleared. Starr saw several columns of smoke rising in the distance.

Soon they were riding alongside acres of vegetable gardens in the open land on the west side of the path. On the east side, among the open trees, were pens holding various types of livestock. People working in the gardens and tending the stock waved and shouted greetings to them.

Farther along, among the largest trees, were the houses made of logs. They brought to mind the American frontier she had seen pictures of in her research. The first sound Starr heard as they rode into Haven was the laughter of the children.

Chapter Eight

Haven

David brought Starr straight to his parents' house, which lay on the outer rim of the village itself. "We're in time for the evening meal," he said, then guiding her up the steps that led to the snug log cabin. He opened the door calling out, "We have a visitor."

As the family came to greet them, Starr felt very strange, almost as if she had been plucked out of the world she had known and set down in another. A tall, sturdy man of about 50 dressed in rough work clothing came to stand before them, observing Starr with dark blue eyes. There was a light in them she had seen before, in much different surroundings.

"This is Starr Omega," David said. "She's come to do a study of the Fields. I thought it would be good if she stayed with us for a time."

"We're glad to have you in our home, Daughter."

Starr looked up at Caleb, David's father,

who was smiling as he spoke, and the formal greeting she intended did not come. The word "daughter" caught at her somehow, and she faltered slightly before answering. "Why, thank you —" she began, but could not find the right words. She had learned how to handle social communication in the City, but there was something in Caleb's smile and eyes that made those skills obsolete.

The woman by Caleb's side saw her embarrassment, and said, "I'm Sarah, my dear. You've had a hard trip. Let me show you where you'll sleep. David, get her things." She led Starr up a set of steep, ladder-like stairs, talking cheerfully all the way. When Starr stepped into the room, she found herself facing a young woman. "This is our daughter, Miriam," Sarah said, then gave Starr's name to the girl. "Supper will be ready in half an hour."

As her mother left, Miriam said, "It's good to have you, Starr. Why don't you take that bed by the window. You can put your things in that chest."

"I hope I won't be a bother, Miriam," Starr said uncertainly. She looked around, adding, "This is a wonderful room." It was a large room, with a ceiling that sloped overhead. Everything was wood, the rafters, the floor, the walls, and the smell of the wood was pleasant. Walking over to the window to look outside, she was delighted to see a pasture where a large number of black and white cows grazed lazily, the wobbly-legged calves staggering after their mothers. They were in a valley, and the dark green outlines of the mountains framed the scene. Turning to face the girl who was watching her, she said, "I don't think I'll be here very long. I'll try not to be a bother."

Miriam smiled, and as she did, Starr saw the strong family resemblance she had to David. She was tall, willowy, and her lustrous black hair fell past her shoulders to her waist. There was a directness in her gaze and a quiet peace in her demeanor that gave Starr confidence. She thought *This will be helpful. I can get close to this girl — get a genuine reading on the mentality of these people.* Aloud she said, "I owe your brother my life. . . ."

Miriam listened as Starr related her narrow escape from the man called Abbadon, then nodded. "You were fortunate. He's a very wicked man."

"I was very lucky David came when he did."

A slight smile tugged at the edges of Miriam's lips. "Lucky? I think it was more than that, Starr." At that moment David called from downstairs, and Starr went quickly to get her things. When she returned and was putting them away in the small chest, she asked, "What did you mean by David's coming when he did was more than lucky?"

Miriam was standing by the window looking out on the fields. When she turned, a slanting ray of sunshine fell across her, heightening the planes of her face. There was a moment's pause as she seemed to think about the question, then she turned to face Starr. "I mean that life is more than just good luck or bad luck," she said quietly. "When people play cards I suppose it's just chance or luck that gives them a good hand. But life's not a game of cards. When something happens to us, there's meaning in it."

Her calm assurance piqued Starr, for it had long been her conviction that life was a matter of chance. Closing the drawer, she sat down on the narrow bed, then remarked, "I believe in cause-and-effect, Miriam. If I'm standing on the edge of a cliff and choose to jump off, I'll be killed. But the choice lies in my hands. I'm responsible for what I do. I guess," she said slowly, "I believe every person makes his own life by what he chooses to do."

"But you didn't *decide* to be killed by Abbadon, did you? It was something beyond your choice, wasn't it?"

"Well — that's true, of course, but —"

"Do you know how large the Fields are,

Starr? And do you know that David doesn't hunt in that area at this time of the year? Just think for one moment, here you are in a place where my brother wouldn't ordinarily be, and at the exact instant when that beast attacked you, he suddenly steps in to save you. What if he had been a mile away, or half a mile? What if he had stopped to cook a meal so that he arrived half an hour later? What if it had been a man less strong than David who could never have defeated Abbadon?" Miriam shook her head, saying quietly, "Some say it takes too much faith to believe that things happen to us for a purpose. But to me, it takes too much faith to believe that the 'right' man just 'happened' to be at the spot at exactly the 'right' time to save your life."

Starr stared at Miriam, trying to find a logical answer, but none came. Finally she said, "So you believe we're controlled by some power. That we're nothing but puppets?"

"No, not like that," Miriam said quickly. "But we aren't alone. Our lives have meaning. I heard of one scientist who said, 'Life is like an onion. You peel it off layer after layer, and when you get to the center — there's nothing there!' I don't believe that."

Starr asked cautiously, "And this — this power that you believe in, that brought David to save me, I suppose you mean the stars?" In the city practically everyone believed in astronomy. One chose an astronomer with as much care as one chose a personal physician. Starr herself paid little heed to the charts her own astronomer gave her, but most of her friends obeyed the signs rigidly.

Miriam gave her an even look, saying, "No. The stars were *created.* How can anything that is itself created have power over human life? But the One who made the stars, He's the one who brought David to help you, Starr Omega." Then she turned, saying, "I'll call you when the meal is ready."

For a few minutes Starr sat on the bed, thinking of what Miriam had said. How was it? *The One who made the stars, He's the one who brought David to help you.* Such a thing went against everything she had been taught all her life, and she rose swiftly in an effort to shake off the effect of Miriam's words. Removing her clothing, she was delighted to find a shower in the small bathroom that adjoined the room. There was no hot water, but she delighted in the cool water that ran down her body, and seemed to sluice away not only the dust and sweat but also some of the tension that had been building up in her since she had left the city.

After drying off with a rough towel, she put on clean clothing: fresh underwear, a pair of denim jodphurs, a red-and-white plaid shirt and clean white boot socks. Her boots were muddy, but would have to do until she had time to clean them, and just as she slipped them on, she heard Miriam call. "Starr? Supper's ready."

"I'll be right there!" she called, then paused just long enough to give her thick hair a quick brushing. Her cheeks, she noticed, were already filled with color from the short time she'd had under the sun, and she saw that excitement brightened her eyes. *You're on your way!* she said silently, then descended the stairway quickly.

Miriam was waiting for her, a smile on her face. "Come along and meet your family, Starr." Her choice of words gave Starr the same peculiar feeling she'd had when Caleb had called her "daughter," and she wondered why such a thing should be. In the City, no one called another "daughter," and the word "family" had become almost archaic.

Entering a large room on the left of the hall, she found herself at a long pine table loaded with food. The fragrance of fresh bread had come to her on her way down, and she stared in unbelief at the plates of meat, the bowls of steaming vegetables and the platters with fresh brown bread.

"This is Starr Omega," Caleb announced.

"She'll be with the family for a time. Starr, these are our sons, Joshua and Timothy — who like to be called Josh and Tim."

"I'm glad to meet you both." Josh, she saw, was much like his mother, having her small frame and brown eyes. Then she turned to Tim and shock ran along her nerves. *He has Down's Syndrome! And he must be at least twelve or thirteen.* She managed to smile pleasantly as they sat down, but the shock persisted. She had never seen a child as old as Tim with Down's Syndrome. In the City they were Relieved as soon as a final diagnosis was made. There had been cases, of course, of mothers hiding their babies for a time, but they were always discovered. Starr suddenly remembered her instructor, the Chief Reliever, speaking on this subject. *You will encounter resistance from parents, but you must get across to them that it is much kinder to Relieve the child at the earliest possible age. In the end the child must die, and it is better for the state, the parent and the afflicted one for Relief to come as quickly as possible.*

"Sit down, all of you," Sarah said, her voice breaking into Starr's thoughts. Starr took the place between David and Miriam, and when they were all seated, Caleb said, "Tim, would you like to give thanks for the food?"

"Yes!" A glow of pleasure came to the boy's

face, and at once he bowed his head. Starr gazed around the table, startled and confused. Then David turned to smile at her, at the same time holding out his hand. She took it without thinking, noting that the others were all joining hands. She reached out with her left hand which was clasped firmly by Miriam. Tim said in a thin, clear tenor, "Thank you God for this food. . . ."

As the boy spoke, Starr was acutely conscious of the intimacy of the moment. Both her hands were held firmly in the warm, firm hands of David and Miriam, and she seemed suddenly to become a part of the small group. She had eaten with groups all her life, but always before there was in her the sense of firm isolation. *You are there — and I am here.* Now the sense of *otherness* had somehow been dissolved. She felt that those in the circle around the table had absorbed her. She was bound together with them by living flesh — and something more than that which she could not identify.

And it frightened her! Starr had carried on a secret battle all her life to maintain a sense of self. Even as a child she had been aware that almost everything in her world sought to draw her into the whole. She had read once an old fable written by one of the OldAge writers, a man called Bram Stoker. It had dealt with vampires, horrible creatures that lived off the life blood of others. And once someone was bitten by one of these, he also became one of the "living dead," as Bram Stoker, the author, called them.

The tale had almost crippled Starr, and for years she felt that something was trying to get at her, to suck her life away — and she suffered nightmares that she had lost herself, had become one of the living dead. But in later years, she had somehow come to realize that the "vampires" of her world were not bat-like creatures as in the tale, but were the agents of her society. The government itself was some sort of life-draining creature, always seeking to break down the secret place in her where that part of her that was *other* than all others were, and to destroy it, or rather, to *absorb* her into itself.

It was this constant attempt of her world to destroy her secret self, to make her part of the whole, that had driven her to take every means at her command to avoid such a fate. She had retreated to her own apartment as a haven from the pressures which, like a mighty maelstrom, sought to suck her into the maw of the City. Her resistance to forming any sort of close personal relationship with others had its source in this fear of being lost, of becoming just a number in a system. Yet, ironically, her loyalty to the City was unabated, for it was the only life she knew.

And now, as she sat there holding hands and listening to the boy give thanks to his God, she was shocked to realize that she *was* being drawn into something! Something in the intimacy of the moment, the warm atmosphere of the small group pulled at her. And for the first time in years she felt a strong desire to let down her resistance, to let herself flow into whatever it was that held this family together.

But she had a lifetime habit of resistance to such things, having seen that the reason people wanted you to "join" them was that they wanted something you had. There was something predatory in the groups she had been exposed to, as well as in the graspings of Sheila Phi, who wanted Starr to satisfy her own greedy hungers.

Suddenly she realized that in her struggle to resist the force that drew her toward those at the table, she had tightened her grip — so much that when Tim said, "Amen!" she opened her eyes to see David casting a glance at her, and was aware that her wrists and fingers were weak with the strain of the effort she had put into them. She glanced quickly at Miriam, and saw both surprise and a trace of pity in the girl's dark blue eyes. Quickly she took her hands away, shocked at the intensity of the emotion that had seized her.

"Well, now, everyone eat up," Caleb said loudly. He winked at the rest of them saying, "Your mother's a terrible cook, but no matter! We'll make the best of it." Picking up a long knife, he skillfully cut off a large slice of meat, saying, to Starr with a gleam of humor in his dark eyes, "Let's have your plate, Daughter, before these gluttons eat everything!"

Starr stared at the huge slice of venison, thinking, *That's one person's ration of meat for a week!* David began spooning vegetables on her plate, making a mound of steamy potatoes, tender carrots, and green peas. She took the slice of fresh bread from the platter, and dipped into a large bowl of creamy yellow butter — real butter, which she had never tasted in her entire life. *I'll never be able to eat that synthetic butter again!* she thought.

The talk ran around the table, mostly about the work that had gone on that day. No one pressured Starr to enter in — she didn't know that Caleb had warned the others, "Let the girl talk when she pleases. Don't pester her!" She listened with interest as Josh told how one of the cows had produced a new calf, and Tim related how he had found 19 guinea eggs that morning — "More than anyone ever found before!" he said proudly.

The food seemed to melt in Starr's mouth, and finally she had to say, "No thank you," when Miriam urged her to try the yellow squash. "I've never had such a wonderful meal!"

"You can't quit now!" Caleb said. "Sarah, didn't you open a jar of those blackberries we canned last summer? I do believe I saw a cobbler on the stove, didn't I?"

"Yes, indeed!" Sarah rose and went to the kitchen, coming back at once with a large bowl which she set on the table. Taking a spoon, she broke the firm brown crust, and steam rose, a delicious fragrance filling the room. "Company first," she announced firmly as Josh reached for the first bowl she filled to the brim with berries and crust. "Put a little cream on that, dear," she said to Starr.

The cream was so thick it wouldn't pour, so Starr ladled out a dollop with a spoon over the succulent berries. She took a tentative bite, and her eyes flew open. It was the first natural fruit she'd ever tasted, totally unlike the hard little affairs from the City Greenhouse that passed for berries. "Oh, this is wonderful!" she exclaimed.

Tim laughed at her. "You've got a white moustache!" he said.

"Never mind what she's got, Tim," Sarah

said. “If I know you, you’ll have cream up to your eyebrows before you’re finished!”

After dessert was finished, Caleb said, “A special treat tonight. Coffee for all.” As his wife brought an ancient enameled pot and filled their mugs with the rich, black coffee, Caleb said, “Everything else on the table came from our own land. But coffee won’t grow in this country.”

To Starr his statement was a revelation. To think of growing all the food she had seen on the large table! The thought of the green globs of paste-like substance she’d forced herself to eat so often at restaurants in the City came to her, and she thought, *It would have been better if I’d never had this meal. I’ll remember it every time I have to eat that tasteless food back home!*

After the meal, Starr insisted on helping with the clean up, and her offer was taken without any protests. “I’ll do the washing if you’ll do the drying,” she said, and soon she and Miriam were busy with the dishes. There was something about working with the young woman that broke down inhibitions, and by the time they were finished and went into the large den, Starr felt very comfortable with her.

David and his father were arguing about the best way to clear new land for the crops. It was, Starr noted, a very amiable argument, for though each man was certain his method was right, each would listen as the other presented his case.

"That's enough talk about your old crops," Sarah said firmly. "You'll bore Starr to death with all that!"

Caleb said at once, "Right you are, wife!" The whole family, Starr saw, had gravitated into the room, and now Caleb asked, "Well, what shall it be?"

"Games!" Tim shouted at once, and everyone laughed. Caleb said, "You and your games! Well, just one or two."

At once Tim chose a simple game that involved thinking of items of the same sort that began with certain letters. He explained it to Starr in excited tones. "I say 'The minister's cat is a curious cat' while we clap our hands, and then you have to call the cat something that begins with a *c*."

"You mean," Starr smiled at the boy's excitement, "I would say, 'The minister's cat is a clever cat?' "

"That's right!" Tim nodded furiously, "You're real smart! Now, I'll start it off, and you can be after me, Starr!"

Tim began the simple rhythm game, Starr making notes in her mind that she would later put into her computer. *Simple people. Play a game that goes far back into antiquity: Check origin of game — was played as far back as 18th century, OldAge.*

For nearly an hour, the family played the games that Tim chose, then Josh groaned, "Enough games for tonight. Let's have some music!"

"Good!" Caleb said, and going to a cabinet on the wall, reached inside and brought back two instruments — one was a violin and the other a stringed instrument that struck a chord of memory in Starr's mind. As Josh took it and placed it in front of him, then took out two round sticks with felled knobs, it came to her: *A dulcimer! Used by mountain people before the end.* She watched with interest as the two began tuning their instruments. Finally Caleb nodded, "Here we go, Josh." He began playing a lively tune on the violin, and the clear notes from the dulcimer made a lovely harmony. It was a rollicking song, and at once the others joined in singing. Their voices blended, and Starr sat back, delighted with the performance. They sang song after song, and clearly they had spent many evenings gathered in this room singing the songs.

Sarah was the best singer, her clear contralto rising sweetly to fill the cabin. When the jolly songs had been sung, Tim said, "Sing the one about the poor man, Mother!"

"No, it's too sad!" Josh protested, but Sarah gave Tim a smile and a wink. Then she lifted her voice and sang a song that was, Starr realized, older than any of her research. Caleb and Josh did not accompany her, and as the woman's voice filled the room, Starr was amazed at the pathos and emotion that Sarah put into the ancient ballad:

"In Scarlet town, where I was born,
There was a fair maid dwelling,
Made every youth cry *Well-a-way!*
Her name was Barbara Allan."

All in the merry month of May,
When green buds they were swelling,
Young Jemmy Grove on his death-bed
  lay,
For love of Barbara Allan.

"Oh, 'tis I'm sick, and very, very sick,
And 'tis a'for Barbara Allan;"
"O the better for me ye's never be
Tho your heart's blood were spilling."

He turned his face unto the wall,
And death was with him dealing:
"Adieu, adieu, my dear friends all,
And be kind to Barbara Allan."

And slowly, slowly rose she up,
And slowly, slowly left him,

And sighing said she could not stay,
Since death of life had reft him.

"O mother, mother, make my bed!
O make it saft and narrow!
Since my love died for me today,
I'll die for him tomorrow."

As the sweet voice reached the end then paused, a silence fell over the room. And Starr felt the power of the song in a way she had never felt before over any music. Tears welled up into her eyes, and she blinked them away angrily, not understanding the sadness that had come to her. The music of the City was composed by computers, and had become so sophisticated that one could buy a program, feed information into it by simple commands:

THEME — SEX IS WONDERFUL;
STYLE — WALTZ TIME;
INSTRUMENTS — STRINGS.

The computer would whirr briefly, then out would come a song exactly as prescribed by the commands.

But none of the songs from her computer had stirred Starr as this one had, and she knew instinctively that this was music that had come out of life! Somewhere in the dim, distant past, there had been a young man who had died for love of his sweetheart. All was lost over the years, except the scrap of a ballad, but that scrap had made the tragedy last for generations.

"Why are you crying, Starr?" Tim asked suddenly.

"Oh, I don't know, Tim," Starr tried to smile. "I guess I just feel sorry for the young man."

"So do I!" Tim nodded seriously. Then he asked, "Father, why do I like a sad song?"

Caleb looked fondly at the boy, thinking about the question. "I think we'd get tired of nothing but happy songs, Son," he said. "Life has its sorrows, you know. And we need to be reminded that they're going to come to us."

"Well, *I* don't have any sorrows!" Tim smiled happily, looking around the room. "We have a good time, don't we, Mother?"

Starr did not miss the quick glance that Sarah threw at her husband, and was certain that the mother's eyes were misty with unshed tears, but she covered it up by laughing and getting to her feet. "Yes, thank the Lord, we do have a good time. Now, it's time for a little from the Book, and then to bed."

Starr saw that they were all looking to the father, and she watched as he opened a thick book that Sarah passed him from a shelf containing a few books. This was not a book in the ordinary sense, but a loose-leaf notebook, thick with well worn pages. Caleb shuffled through the pages, found a place, then began to read:

"Though I speak with the tongues of men and of angels, and have not love, I am become as a sounding brass or a tinkling cymbal. . . ."

He was, Star realized, a fine reader, his voice deep and clear, and it was also evident that he knew the passage he read by heart, for he would often look around the room without missing a word.

And it was a beautiful passage, though Starr did not recognize it. It spoke of how love was the most important thing in all the world, and then it listed the characteristics of genuine love: how that it was never jealous and always rejoiced when others were blessed. And love never grew angry or bitter when a person was treated badly. As Caleb read on, Starr felt a strange longing growing in her heart. It was something that she had felt before to some extent, but now it came almost like a pang. She leaned her head back, eyes closed, and the longing grew until it was a pang sharper than anything she had ever felt.

Yet — she realized that she did not even know what it was that she was longing for! Yet it was there, keener and more demanding than a physical desire for food or drink. She felt like a fool, sitting there almost sick with desire for something she could not even name!

Suddenly a fragment of memory came to her, tantalizing and vague as an almost forgotten tune that slipped around the edges of the mind. What was it? Something she had read — about just such a feeling. She had learned long ago that to throw her mind into a search for such things was useless. Better to relax, to think not of the mystery that lay just outside the rim of her consciousness, but to wait —

And then it came, with a clarity so sharp that she could almost see the black words on the white paper. A seer from England! What was his name? Yes! Lewis, that was it! C.S. Lewis! Now it all came to her, the words of the man so long dead. Words she had read in a book, that told of Lewis' experience with just such a longing as gripped her. He had related how that he had seen a glimpse of beauty while still a boy, that it dominated all his life. He told how he had spent years trying to find that beauty — and had failed completely. He called the emotion, the longing for beauty or truth by its German name, *sehnsucht,* Starr remembered. She let the memory flow into her, then it came to her. Lewis had found that beauty no place but in God! That was the reason, she realized at once, that the works of C. S. Lewis were on the Index, forbidden to publish or to possess! Only her position as remedial historian had brought her to the text of the man, and now she knew exactly what he had felt!

". . . For now we see through a glass, darkly; but then, face to face: now I know in part; but then shall I know even as also I am known. And now abideth faith, hope, love, these three; but the greatest of these is love."

A small chorus of "Amens" went around the room, and without a pause, Caleb began to pray. Starr instinctively bowed her head and closed her eyes. It was a simple prayer, not long. Caleb thanked his God for all the goodness he and his family had received, asked protection from the dangers that might lie ahead — and closed by saying, "Thank you for delivering our guest from harm. Be near to her, Lord, even in her heart." And then he said, "In the name of Jesus Christ we ask these things. Amen!"

Once again the family said "Amen," and then Sarah said, "To bed with you Tim, and the rest of you, as well!"

Starr rose and feeling Caleb's eyes on her, said, "Goodnight. And thank you, for your hospitality."

"God be with you, Daughter," Caleb nodded. And when Miriam had led Starr up the stairs and the others had gone to their own rooms, he said, "There's a child with a heavy spirit, Sarah."

"Yes. But she felt the hand of the Lord tonight. I saw it in her eyes. Come, we'll pray for her. I feel that she has a dark way to tread."

Caleb stared at his wife, for he had long known that she had a sense for such things. The two of them joined hands, and offered a fervent prayer for the young woman who had come into their lives.

Upstairs the two young women got undressed and ready for bed. "Would it bother you if I did a little work before we turn the light out, Miriam?" Starr asked.

"Not at all. I want to read for awhile myself."

Star took out her small journal and yellow pencil. But she soon discovered that she could not begin. Usually words flowed from her, but the day had been so strange that she could not put her feelings into cold words. Shaking her head with a stubborn motion, she determined to write down *facts,* leaving her emotions out of it:

My first contact with Primitive family. They are friendly and hospitable, very much so! Can they be this open with every visitor?

It seems too good to be true. As time goes on, in all probability I will discover they have their angers and jealousies just as other people.

The boy Tim came as a shock. They must know that he is doomed! There is little medical knowledge here, but they must have seen other children afflicted with Down's Syndrome. They are especially gentle and tender with the boy, and he seems happy. But it is a disturbing factor! The State knows best, of course, and has laid out the principles of Relief so that suffering will be alleviated. Yet — they are all so — !

They are extremely religious. And the worst sort of religion, that of the Crossbearer. It came as a shock to hear the head of the family pronounce the name of Jesus quite openly. In the City he would not last long! He might survive in the Fringe, but even there the most flagrant cases of this primitive superstition do not go unnoticed by the leaders in the City!

During the course of the evening, while the father was reading from some ancient text, I was stricken by some sort of emotional disturbance. No doubt it was the result of my terrible fright with the man called Abbadon. I had thought myself more stable than to fall into such a tragic state, for I have always prided myself on keeping my emotions under firm control, but tonight, as Caleb read about love —

Here Starr broke off suddenly, panic stricken, for even the memory of the words Caleb read caused the longing to rise in her breast! She closed the notebook hurriedly, put it in the locked section of her bag, then lay down on the bed, trembling despite herself.

"Good night, Starr," Miriam said quietly, looking up from her book. "Tomorrow I'll show you around the village, if you like."

"Yes, thank you," Starr said quickly. She rolled over, pulling the cover over her, but sleep did not come easily. She lay there until Miriam finished her reading, turned the lamp down, and went to bed. It was very quiet in the attic room, and soon Starr heard the even breathing of the other girl.

Finally she grew calm, and the last thought that came to her before she drifted off was a phrase from Caleb's book: ". . . the greatest of these is love."

Chapter Nine

A Trip To The Market

The soft sound of Miriam's voice drew Starr out of sleep, but she made no sign of awakening. On her first morning, she had been alarmed when the voice of the girl had wakened her. For a moment she had been completely disoriented, not knowing where she was. Tense with fear, she lay there ready to leap out of bed and defend herself — then memory came flooding back, and she glanced across the room to see in the dim light that the girl was kneeling beside her bed.

This was the sixth morning of her visit, and Starr lay quietly under the warm coverlet until Miriam finally rose and slipped quietly from the room. As soon as the door closed, Starr arose, put on the wool robe Miriam had given her, then lit the oil lamp on the table. It cast a golden gleam over the room as she quickly pulled her computer from her pack and began her daily report. Morning, she had discovered, was the best time for this. The women were preparing breakfast and the men were taking care of early chores. The days were so full that by bedtime she was ready to go to sleep at once. She had discovered that the fresh air, vigorous exercise, and delicious food were better for making one sleep than the Morpheus Capsules she had brought along in her kit. Nor had she needed any other of the drugs she had brought, not even the Uplifter pill that she had assumed was necessary each day to get her to a working pitch. For most of her life Starr had taken pills to sleep, to heighten her mental processes, and to blot out the fears and anxieties that lay under the surface of her consciousness.

Sitting down on the bed with the Voice-Writer in her hand, she was struck by the thought *Why don't I need all the drugs I've used for years?* It was a disturbing question, somehow, and she shook it off quickly as she began speaking into the VoiceWriter.

"Sixth day of Starr Omega Investigation. Subject: Familial patterns of Primitives."

"Most common social structure in the village called Haven is the nuclear family, consisting of father, mother and one or more children. Basic power flow begins with father, the strong authority. Archaic submission patterns exist in wife and children. According to all modern sociological research, such a pattern of male dominance should produce extreme tensions. Even in OldWorld's primitive structures, toward the end progressive groups, such as Women's Lib, Gay Power and the ACLU, had managed to eradicate the older religious and mythological concepts of marriage. It is to be noted that the misery and unhappiness that have proven to result from such archaic structures are well hidden by the dwellers of the Fields. . . ."

Starr paused, shut the VoiceWriter off, and picked up her notebook. Slowly she wrote: *Why are they so happy? I know it must be some sort of self-hypnosis! They do everything that our finest scholars and social workers have proven to be evil — yet they seem so content!* She jammed her pencil so hard on the exclamation point that it snapped, and with a gesture of frustration, she suddenly thrust the tablet and pencil back into the kit, shoved it into a drawer, then began to get ready for breakfast. But as she brushed her hair, the enigma of Caleb and his family kept gnawing at her.

There was, for example, the matter of Caleb's father, Amos. Ever since she had been taken to meet him on the second day of her visit, she had been haunted by his face. Caleb had said cheerfully to her after supper, "I'd like you to meet my father, Starr." She had agreed and followed him upstairs. "He's feeling very well today," Caleb had informed her, opening the door, and following her into the small room with a tall window that looked down onto the front yard. The sunlight had blinded her momentarily, and she had stood there blinking as Caleb had said, "Well, now, Father, you have a visitor! This is Starr."

As her eyes grew accustomed to the glare of sunlight, Starr had looked for Caleb's father, but no one was in the room — at least not that she had seen at first. Then she realized that a man was in the bed beside the window, a small man, shrunken by age until he made only a slight outline under the cover. His face was shrunken as well, so that there was a skull-like quality about it, but the eyes that looked at her were alive. His lips were thin and seamed, and when he spoke Starr had to strain to catch his words. "Welcome . . . !"

Starr had swallowed, then nodded, forcing herself to speak normally. "I'm glad to meet you, sir." That was all she had been able to manage, but Caleb had kept the conversation going, telling his father what Starr had been doing, throwing in bits of news about the farm and the family. As he had talked, Starr had sat there, looking at the ancient face and slowly gotten over her shock. The man was the oldest human being she had ever seen, for in her world no one would be permitted to live so long past usefulness. As a Reliever she had been assigned to many who were no longer able to maintain their level of usefulness; some of them had been rather painful cases. One woman came to her mind, a middle-aged woman who had been signed over to the Department of Concern by her son. *She's not able to keep up any more* he had informed Starr, and the mother's eyes had dulled as she was told that her family had assigned her to be Relieved.

And yet, Starr remembered, that woman had been much stronger than Caleb's father! What amazed her most, however, was the gentle care and love that the old man received. He was almost helpless most of the time, and the family all took part in caring for him. He was, she found, much like a new infant, having to be bathed and fed special food. Yet there was no complaint from any of them, even Josh, who was very impatient with most things.

After that first visit, Starr had become intrigued with the situation, for it was something totally foreign to her experience. She visited with the old man more than once. Sometimes he just slept, but at times he became almost animated, speaking of the old days. She had heard that it had been that way with elderly people before they were Relieved

— that the days long gone were more real and vivid to them than the present. David had spoken of that once, as the two of them had sat beside Amos. "He can remember almost everything that happened thirty or forty years ago," he had said thoughtfully, "but he sometimes can't remember anything about what happened this morning."

As Starr finished brushing her hair, she suddenly thought, "I wonder if he could have known my parents?" The thought came unexpectedly, and for a few seconds, she sat there staring in the mirror. She had found herself thinking more about her parents lately, perhaps because she knew that they had been sentenced to the Fields. Always she had tried not to think of them. A Psychologist had warned her when she was only a child, "It is not healthy for you to think of them. They were Heretics and the City became your family to save you from them. Banish them from your mind. If you cannot do so, we have medication to remediate the problem."

Starr had resisted medication, suppressing all thoughts of her parents when the Examiner tested her at regular intervals for soundness of mind. But now, the City seemed far away. Almost every night since she had been with David's family, seeing the closeness of the father and mother with their children,

Starr's long-buried memories had begun to surface. Sometimes it was in the form of dreams which came just before dawn. The dreams were frightening, yet strangely alluring. In one of them she saw herself as a little girl sitting on the shoulders of a tall man with black hair. There was a woman by his side. The three of them were walking beside a small stream. The man had taken her from his shoulders, and the woman had removed her shoes and stockings so she could wade in the pool. The water had been cool. In the dream all of them had laughed when Starr almost fell into the water. She could remember few details, but the man had a kind face and a black patch over one eye. He had a terrible scar on the right side of his face. The woman was beautiful, and in some of the dreams she sang a song to Starr — something about a lamb. One thing that made this dream somewhat frightening to Starr was the fact that she wore something that had belonged to her parents — a small gold pin in the shape of a lamb. She had not worn it for years, knowing that if anyone knew of it, she would not be allowed to keep it.

Suddenly, she reached back into the kit, removed a small leather case and drew from it the tiny pin. It was a simple piece of jewelry, of little value. But as she held it she realized for the first time how much she treasured it. It was the only link to her past — to her parents. With a defiant gesture she suddenly pinned the brooch over her left breast. She had worn it a few times, when she was a child. But it was always when she was alone. Now, as a woman, she wore it for the first time. It seemed to glow in the rosy light of dawn that pierced the window.

Starr left the room, her face intent as a vague plan began to form in her mind. By the time she got downstairs, she knew what she would do. Saying nothing to any of the family, Starr smiled and spoke cheerfully to the others. When the meal was over she said, "Let me take your father's breakfast to him this morning, Caleb."

Caleb looked at her with surprise, saying, "Well, he's not a very neat eater, you understand."

"Oh, that doesn't matter," she answered quickly. "I want to help, and I'm not as quick as Miriam. If he's feeling well, I'll sit and talk with him awhile." She quickly made the fine mush with the butter and a little salt as she'd seen Miriam do several times. Then she filled a cup with hot milk and made her way back to Amos' room.

"Good morning," she said cheerfully as she entered. "I'm your new maid, Amos. Let me help you sit up. . . ." She saw at once that he was alert, the old eyes bright as a bird's. She kept up a busy chatter while she sat beside him and fed him small portions from a pewter spoon. After he had eaten all he wanted, she cleaned his face, put the tray aside, and asked, "May I visit with you awhile?"

"Yes, of course," he nodded, his thin lips trembling slightly. "You are —" he said, then his eyes clouded. "I forget — !"

"I'm Starr," she said, and waited until his eyes gleamed with recognition. "I've been thinking of the past," she said casually. "What it must have been like in those days. I suppose it was very hard?"

Her question caught his interest, and at once he began speaking of the past. "Yes, it was hard," he nodded, "There was nothing easy about the Fields when the first of us came. The land had to be cleared by oxen. At first we had to live in caves. . . ." He rambled on, sometimes getting the present and the past confused, but out of the patchwork recitation a picture began to emerge. Starr had always had an analytic mind, but at the same time was able to see things in vivid images. As the old man spoke, she saw a ragged group of outcasts cut off from the civilized world and forced to struggle mightily for bare survival in a hostile and dangerous world.

Starr leaned forward, her eyes wide as Amos spoke of the fierce battles with wild animals that had begun to proliferate as the earth renewed itself. She was horrified as he spoke casually of one of his children being dragged off by a band of marauding wolves, the child's frantic screams being heard throughout the night. "There was nothing we could do," Amos said slowly, his eyes turned bitter over the tragedy that had taken place long before Starr was born. "If I had gone outside I might have been killed as well. Then the others would have fallen to the pack." He stared at her, his eyes opaque, yet filled with grief. Then he smiled suddenly, his entire countenance lighting up. "But God was faithful," he whispered. "He brought us through the flood and through the fire. Blessed be the name of our God!"

Once again the longing that had been coming to Starr flooded her. She could do nothing but sit and listen as Amos spoke of the goodness of his God. His simple faith was alien to all that she had acccptcd as true and, though she let none of it show in her eyes, conflict in her breast raged.

Finally, seeing that the old man was tiring Starr asked, "Did you ever know a man with one eye and a terrible scar on his face who was exiled from the City? It must have been about twenty years ago? He and his wife?"

"There were so many," Amos muttered, then he cocked his head to one side, a thought coming to him. "But — yes! There *was* such a man!"

Starr said nothing, for she had learned that to press people was to interfere with their memory. Patiently she sat there, while he searched through his past. He was like a shopper, going through a large box filled with all sorts of things, looking for one specific item, and discarding all the rest. "He was a very large man with reddish hair — no, wait, that was Thaddeus, not him. But he did have one eye, I remember that very well!" He nodded emphatically, demanding, "Do you know how I can remember that? Because he was such a strong fellow! Why, he had more strength than three men!"

"Really? What was his name?"

Amos stared at her, his eyes going blank. "Name? Why, I can't recall. Was it James something? No, not that. An unusual name — never heard it before."

Starr asked casually, "Was it something like Jason?"

"Jason!" A light burned in the old man's eyes, and he nodded proudly. "Yes! That was it — Jason!" He peered at her curiously. "You couldn't have known him. You're too young."

Starr said tightly, "He — I think he was my father." Then she had to ask, "Is he still living?"

"Oh, my, I have no idea!"

"Do you know where I might find him? Or someone who could tell me about him and his wife?"

"No! No! You don't want to be looking into that!"

"But why not?"

"Because he was sent to the Badlands, him and his wife!" The thought seemed to distress the old man, and he slipped down in the bed, closing his eyes.

Starr leaned forward, knowing that she might never get another chance to discover anything from Amos. He had been more lucid than she had ever seen him, and it was unlikely that he would even remember what he had told her. "Please! Help me, Amos! I must find them!"

The ancient eyes opened and she saw that he was struggling with something. Finally he whispered so faintly that she had to put her ear almost to his lips to catch his words: "Assad! He will know!"

"Assad?" she asked, but he had dropped off abruptly in the way of the very old and the very young. She pulled the cover over Amos, stood to her feet and whispered, "Assad!"

Then she picked up the tray and left the room.

"You stayed a long time," Miriam said as she entered the kitchen. "Was he awake much?"

"Yes. He was telling me about the old days."

"He's one of the few who remember them. David and I would like to make a book from what he's told us."

Starr exclaimed, "Oh, I'd love to read it, Miriam!"

"Of course. It's all in my sorry handwriting — mine and David's. It's in the bookshelf by my bed — the one in a green notebook. Help yourself, Starr."

Just then David came in, "I'm going to the market. Anybody who wants to go better get cracking!"

Starr, urged by Miriam, went outside to find the wagon loaded with vegetables. "Climb aboard, woman," David smiled at her, holding out his hand. "You're like all the rest, I see. Can't resist a trip to shop." His strong hand closed on hers, and she was pulled bodily up into the wagon. She fell against him, then pulled back at once, moving to the far side of the seat. He gave her a quick look, "Where are the others?"

"They're not going. But your mother gave

me a list of things to bring back."

David spoke to the team and they started off at a brisk trot. As they moved down the dusty road David asked, "How was Grandfather this morning?" He listened as she gave her report, then nodded. "I'm glad you spent some time with him. When he is up to it, he likes to talk about the old days."

Starr hesitated, then made a quick decision. "He really was a help to me, David. Part of my job here involves looking up a few people. And your grandfather gave me a clue about the one I need to see most."

"Oh? Who is it?"

"A man named Jason. He came here about twenty years ago, but your grandfather remembered him." When she saw the name meant nothing to David, she added in a casual tone, "Your grandfather said I could find out where he was now from a man named Assad."

David stiffened and turned to fix her with his eyes. "Assad? I only know of one — and you wouldn't be interested in meeting up with him, Starr."

"Oh, I suppose I'll have to, David. Part of my job." Starr was reluctant to tell her secret to anyone, but she could see that David was going to be difficult. "He's in the Badlands, your grandfather said. I think I'll go over there next week."

"That's not the best idea," David shot back. "Remember your friend Abbadon? That's his home territory. And to give you an idea of what sort of place it is, they consider Abbadon a pansy!"

Starr's heart sank, but she said stubbornly, "I have clearance to go to any section of the Fields, David."

"You think a piece of paper means anything to that crowd? Starr, even the Border Guards are afraid to go there. They won't ever go alone. Always take a troop, and be well-armed." He glanced at her, his eyes serious. "They'd take you, use you up, and knock you in the head when they were finished!"

The rough edge of his voice, as well as the raw warning about the place, silenced Starr momentarily. She sat quietly in the seat, noting the rich fields of green grain waving in the breeze. All the time she was aware that no matter what David said she was going to make the attempt to find her parents. Nothing further was said about the matter.

Soon they reached the village square which was humming like a beehive. Starr had been there twice earlier in the week, but it had been almost empty. Now there was a teeming mass of people, all of them seemingly wearing the brightest colors available. The square looked to her like a kaleidoscope with the colorful shirts of the men and the dresses of the women constantly in motion. Booths of every sort were set up and a babble of voices filled the air as the vendors called out loudly to entice customers. David found a place to hitch the team and then helped her down. "Got your list?"

"Yes, but I don't have any of your money."

"You won't need much," he said. "That's why we brought the vegetables."

For the next hour Starr had a wonderful — and confusing time. When she found an article at one of the booths, a large iron pot that Sarah had listed, she was astonished at what followed. Seeing her hesitate, David whispered, "I'll give you a quick lesson in how to get a bargain." He sauntered up to a booth, picked up a pot and looked at it with a frown, then tossed it down saying, "Thaddeus, that's the worst piece of work I've ever seen! I knew your creativity would suffer when your dear old father died!"

The vendor, a chunky black-eyed man with a bristling moustache that seemed to quiver with rage, shouted at him, "You pig-farmer! You wouldn't know good iron work from a potato!" Snatching up the pot, he began pointing out the fine qualities of the piece. When he had finished David said, "I feel sorry for you Thaddeus, but because I liked your father and promised him to try and keep you from starving, I'll give you two bushels of the finest, sweetest carrots ever dug."

The moustache twitched and Thaddeus laughed scornfully, "You can't get rid of those stringy old carrots, so you want my family to try to digest them, is that it? Two bushels, ha!" He pulled at his hair, "David, because you're young and not likely to do well with that sorry farm, I'm going to do you the favor of letting you take this splendid, hand-crafted piece of work home with you for your dear mother — for only ten bushels of your withered carrots."

Starr stood there amazed as the two fought back and forth. She had never bargained for anything in her life. When she wanted something, if she had the Credits she would simply thrust her hand out, allowed the machine to read her number — that was it. After what seemed like a long and angry battle between the two men, David finally agreed to deliver seven bushels of carrots to the home of Thaddeus in exchange for the pot.

"See you at the Ecclesia in the morning, Thad," David said with a smile as he tucked the pot under his arm.

"Of course, and give my best to your family, David," the vendor smiled at both of them, then turned away to begin shouting at another customer.

"Now, you can do the rest of the shopping," David said with a faint smile. "Here's a list of the stuff in the wagon. Don't let yourself get cheated."

Starr opened her mouth, but before she could object, David turned and disappeared into the milling crowd. Feeling helpless and afraid, Starr wandered around the market. Finally she said to herself, *This won't do! What are you afraid of? Just march over there and start shouting!*

And it worked! She haggled with a young woman who almost screamed when she made an offer of five ripe melons for a bright silk scarf. But Starr held her own and got the scarf for only six melons! She had no idea if she had gotten a bargain, but it had been exciting. She lost herself for the next three hours, bidding, arguing, walking away then allowing herself to be called back.

David found Starr, and after looking at the pile of goods he whistled, "What a stack of plunder! You didn't get all this for that wagon load of vegetables?"

"Yes! It was fun, David! I've never had such a good time!"

David looked down at her, noting the color in her cheeks, the brightness of her gray eyes. She was wearing one of Miriam's dresses, and it was the first time he had seen her in anything other than jeans and boots. The dress was a bright green, with a white bodice, and, since she was larger than Miriam, it fit her snugly at the hips and breast. He wanted to say, "You look beautiful." But instead, he held the words back, "If a visit to a market is the best time you've ever had, you must have had a real dull life, Starr."

Starr saw admiration in David's eyes. It was what she had seen in the eyes of many men, and learned to dread it. But Starr wanted to please him. When he said nothing except, "Let's walk around for awhile," she was disappointed.

They spent another hour at the market, then David delivered the vegetables to the different vendors. They were about to get in the wagon, when a man came running up to say, "David, tell Miriam I'll be by tonight."

He stared at Starr, "That's Saul Thomas. He wants to marry my sister."

Starr looked surprised. "He's much older than she is, isn't he?"

"Sure. He's got a lot of competition, but he's the richest of all her suitors." The thought somehow displeased him and he said, "He's one of the leaders at the Ecclesia. That means a lot."

"I don't think I understand the Ecclesia.

What is it, David?"

"It's those of us who follow Jesus, Starr." He saw her look of perplexity, "You'll know more about it after tomorrow. That's when we meet. Will you come?"

Starr hesitated, aware that two forces were at work in her. Part of her wanted to go, the other part didn't. But it was for this she had come. So, she nodded, saying, "I'd like to go very much."

"Fine." They came to a shallow creek a few minutes later, and he stopped to give the horses a rest and allow them to drink. Tall trees shaded the crossing and as the horses snorted and pawed the water, they sat there until he turned to her. "You know something?"

"What?"

"You look very nice today."

The compliment caught Starr off-guard. Her cheeks suddenly grew warm. She lifted a hand to one to conceal the color, "Thank you."

"As a matter of fact, you look so nice that I'm going to break a promise I made to my mother."

Starr stared at him in bewilderment.

"I promised mother I'd never kiss a young woman on our first buggy ride," David continued. "But you look so fresh and pretty, I'm going to make an exception."

David pulled her close, and before she could resist, placed his lips firmly on hers. Starr found herself responding to the kiss, adding her own pressure. As the kiss continued, she found herself disquieted at the feelings that rose within her.

Finally, David drew back, and a smile pulled at the corners of his lips. "You mustn't expect such favors *every* time I take you to town, Starr!"

Then Starr realized that David was teasing her, and pulled back. She tried to be angry, but the sight of his amused smile made that impossible. She laughed suddenly, determined not to let him get the better of her. "I can see you've had lots of practice. But I'll take the matter up with Sarah."

David became uneasy, "I — I don't think that would be such a good idea. Mother is — a little old-fashioned."

He suddenly looked like a young boy caught with his hand in the cookie jar. This delighted Starr. "Yes, I think your mother should know what her son is up to. I'll talk to her as soon as we get home."

David looked uneasy, then he caught a glimpse of the smile that was pulling at her lips. "Well, if I'm going to be whipped, I might as well give her a good reason!"

He reached for Starr as she drew back

laughing. “No, I won’t be a party to corrupting your promises to your mother. Now, get those horses moving!”

The incident had lightened their moods, and they laughed at little things on the rest of the trip. But just as they pulled into the yard, Starr said, “David, I enjoyed the day.” She was feeling happier than she could remember, and impulsively leaned over and kissed his cheek. “Thank you, David!”

He stared at her, touched his lips, then said with a gleam of humor in his dark eyes, “Well, let’s go give Mother the bad news!”

“Maybe we’ll wait until next market day,” she said, smiling up at him. Then they both laughed and got out of the wagon. Starr felt secure in some strange way and the thought of his strength gave her a strong pleasure — which for some reason, disturbed her.

Chapter Ten

Light Of The World

Starr sat on a fallen tree high above the River looking out over the dark expanse of water toward the light-rimmed east. As the thin glowing band grew wider and brighter it seemed to her that she was inside some monstrous beast whose jaws were being pried apart to let the light shine in darkness.

A different dream had come to her during the night and she awakened with a terrible longing, an aching deep inside her, for which there seemed to be no cause nor cure. She could remember little of the dream (wandering in a place shrouded in heavy fog; circles of light with hands reaching out to her that vanished as she tried to grasp them) but it had left its mark.

The morning wind was sighing through the tops of the tall pines on the bluff. Starr drew the light blanket around her against the chill and watched the glow in the eastern sky change from red to a pale pink. The stars slowly faded as the sun slipped the bonds of night and rose from the earth, turning the sky white as bone. The surface of the River ruffled in the wind, dancing and sparkling as if to celebrate the new day.

"The Light of the World."

Starr turned quickly around. "David, you startled me!"

"Sorry," he said, sitting beside her. "I was watching the sunrise. Guess I was thinking out loud."

"What was that you said?"

"The Light of the World," he replied. "It's from the Bible. That's one of the many names for Jesus."

Starr watched the sun as it grew brighter; scattered trees laid their long shadows down across the prairie. As light touched the tall grass dewdrops glittered like millions of jewels reaching to the far horizon.

"Are you all right?" David asked.

"It's so beautiful," she said, pointing down across the River where light was pouring into the land. "Is that what He's really like, this man you call the Light of the World?"

"In a way He is, but no one can truly describe Him. You have to see Him with your spirit — know Him with your heart."

"It's all so confusing." Starr rose and walked to the edge of the bluff.

David followed and stood beside her,

seeing the breeze catch her hair, swirling it in a dark cloud about her head and shoulders. Her brow was furrowed in thought as David's eyes saw in her the face of a child. "The wise are confounded and the children understand," he said.

"What's that from?" Starr asked, turning to face him.

"From me," he replied. "It's how you come to a knowledge of God's word. More than that," he continued, "it's how you come to know God. Jesus said, if we don't become as little children we can't see the Kingdom of Heaven."

"I'm not going to think about this anymore now," she said, shaking her head slowly. "I'll go insane if I do."

David put his arm around her and they stood together on the high bluff overlooking the River. Behind them were the sounds of the village coming to life and before them lay the paths they would choose as they journeyed through life.

"Well I have to get to work," David said. "You coming?"

"David?"

"Yes."

"David, I have to go look for my parents. I'll never have any peace if I don't do everything I can to find them."

She had spoken of her parents earlier, but David had not expected this. He frowned down at her. "You're going into the Badlands? You wouldn't make it through the first day in that place."

"I don't think I'll make it anywhere if I don't try to find them, David."

He could see there would be no changing her mind. "I'll speak to my father. Arrangements have to be made. We'll leave at sunrise tomorrow."

As Starr came down the stairs she saw David packing his knapsack for their journey. There was smoked meat, dried fruit, brown bread and other supplies. Starr's saddlebags lay unopened on the table.

"Good morning," David greeted her while continuing to pack. "You ready for our big adventure?"

Sleep still clouded Starr's mind. *How can he be so cheerful this early in the morning?* she thought. "I'd better be, hadn't I, since the whole thing was my idea?"

"Here," David said, smiling as Starr rubbed her eyes with both hands, "put some of this stuff in your saddlebags."

Starr packed the remaining supplies and greeted Sarah as she came into the room with a steaming platter of scrambled eggs, smoked ham and hot buttered biscuits. David cleared the table and the three of them sat down to breakfast.

"Where's your husband?" Starr asked.

"Oh, he left hours ago. Had to take some men to begin clearing one of the new fields. It's a long way off. I expect he'll be gone for three or four days," she replied. "I hope you two children won't be gone much longer than that."

David sliced a generous portion of ham and speared it with his fork. "She's like an old mother hen, Starr. Always worried about something happening to one of her brood."

"Oh David, that's not true," Sarah admonished him quickly. "I just like having the family together, that's all."

"Well, you don't have to worry about Starr and me. She's been trained by the Border Guards," David said. "Men like Abbadon tremble at the very mention of her name."

Starr had a mental picture of Abbadon's hulking, smelly presence. He had his knife in one hand and a huge piece of greasy meat in the other. A voice called out, "Starr Omega" and he began trembling uncontrollably, slinking away from the sound of her name.

She smiled slightly, then wider and finally laughed out loud. David and Sarah had joined in, and after a few giggles and chuckles, they sat around the breakfast table smiling and content, and it was as if they had known each other for years.

"You never know what's going to strike somebody as funny, do you?" David asked.

"He doesn't seem nearly so dreadful now," Starr said. "I may actually laugh at that creature if I ever see him again." And she meant it. Somehow laughter had healed the wound Abbadon had inflicted in her soul, and the consuming fear she had carried with her since their encounter had vanished.

"I wouldn't advise that," David warned. "Abbadon isn't known for his sense of humor."

They finished breakfast and went outside to ready the horses for their journey. It was still dark and the full moon bathed them in a gossamer light as it settled in the western sky. They finished their work and mounted the horses just as Sarah came out the back door and headed toward the stables. She walked between the horses and David leaned over and kissed her on the cheek.

Starr was surprised when Sarah turned toward her, but she followed David's example, and kissed his mother on the cheek. How strange and yet how comfortable she was beginning to feel with these people. *And how easily I'm adapting to their customs,* she thought. *Next thing you know they'll have me giving thanks at the table.*

Sarah reached up and took David's hand in her own, then Starr's. "Heavenly Father, we praise you for another day you've given us. Bless these children, protect them on their journey and return them safely to us. We thank you for your angels that will keep them in all their ways."

Starr felt a stirring in the air and a small chill down her back. She quickly opened her eyes and saw the bowed heads of David and Sarah. She felt the warmth of Sarah's hand. Then Starr gazed heavenward at the vast scattering of stars: cold, silent, remote.

Soon David and Starr were riding in moonlight through the pasture where the cattle were still bunched together and the calves close to their mothers. This was the last hour of darkness. Then they entered a forest, following a path through the towering pines and underbrush until sunrise when it met the open rolling hill country. Two doves flew by in front of David and Starr, the early sunlight silvering their wings as they sped toward the small stream at the bottom of the hill. "Light of the World" flashed across Starr's mind like the silvered wings of the doves.

All morning they rode with the sun warming their backs. The mountains across the River appeared to be getting smaller and the rolling land they rode through became a dry, rocky plain. At noon they stopped under a solitary tree on the bank of the River to eat their lunch.

"When we cross the River the country and the people will be different than what you've seen so far," David said, eating some of the dried peaches.

"Worse than Abaddon?" Starr asked.

"Probably not, only more of them. I don't want you to ever be separated from me, not even for a short time," David told her solemnly. "You may not think it's important, but take my word for it. Things can happen very quickly out here."

After eating and resting they crossed the River at a ford upstream where the water was swift but deep enough to reach their stirrups. As they rode, Starr noticed a dead tree on the opposite bank. It was wide as a barn and a hundred feet tall — its dead limbs barren and black against the bright sky. Vultures sat silent and still on the highest branches, somewhat like statues adorning an alien house of worship.

David noticed Starr staring up at them. "Our welcoming committee," he said. "They got the job because they're the friendliest looking things in this part of the country."

Starr looked at him with a puzzled expression.

"See that one at the very end of the highest limb," David said, pointing to it. "That one's a prize winner — sweetest smile."

Starr laughed softly. *Humor is an art I must learn. It seems to take the sting out of life.*

David and Starr continued on the north bank of the River until they reached a dam that formed a vast lake. David stopped at a coppice near the shore and dismounted. "We'll spend the night here," he said, unsaddling his horse. "It'll be dark soon."

David took the canteens to the lake and filled them. When he returned Starr was unpacking the food and blankets. "Aren't you going to make a fire?" she asked.

"Not in this country," he answered. "Out here a flame might attract some big, cantankerous moths."

"Moths!" Starr said, then realized what he meant. *I'll catch on eventually I guess.*

At the edge of the trees David and Starr sat on the blanket they had spread over a bed of leaves. As they ate smoked ham and bread they watched the sun as it painted the western sky with streaks of peach, violet and pink. As the sun sank into the lake it seemed to set the horizon aflame and turn the water into blood.

"What was it like out here, David, when

everything was poisoned and dying?" Starr asked. "Does anyone know?"

"Some of the tales have been passed along. They're not very pretty," he said.

"I've studied the history of how it came about: pollution from industry and automobiles; depletion of the Ozone Layer; destruction of the last remaining rain forests. That seemed to be the final blow. Even the scientists leading the reform movements didn't expect what happened. It came too suddenly and the extent of it was far beyond anything they had imagined."

David was looking out over the lake at the last of the light. The first star appeared in the sky. "It all sounds so bare and sterile when you tell it like that; those histories written inside the Domes," he said, turning toward her.

"Tell me the stories of your people, David."

"I'd hear them talking about it as a boy — those old tales that were passed down through the years." David's voice was growing softer like a whispering in the trees. "The thing I remember most was the burning rain. They say when it touched the skin it was like liquid fire and made terrible blisters that wouldn't heal. After it rained, the little sunlight that came through those poison clouds made the trees and plants smoke and burn. The earth was blackened.

"Sometimes the clouds would settle to the earth and wherever that happened the people would . . ." David stopped. The sound of frogs croaking came to them from the lake and the crickets provided counterpoint from the grass along the shore like a miniature string section. "I don't think I want to talk about this anymore, Starr," he said. "I might say some things you wouldn't want to hear."

"No. I want to hear."

"As you know, the New Age or New World Order governments were ruling the country — the earth! When the Domes were built, there was no room for all the people. The helpless were either killed or forced outside to die. Old people, children like Tim, the sick and the cripples were treated worse than animals. And of course the Christians were the first to go." David looked at Starr. "I don't mean to condemn you," he said. "Those were desperate times. I'm sure your government would never condone such things now."

Starr determined then not to talk about the government of the City, especially her occupation. *How could he ever understand the enlightenment of the civilized mind?* "Yes, those were desperate times," she said confidently, but there was a tightness in her chest that she couldn't explain.

"Why don't we talk about something more

pleasant?" David asked, rising from the blankets. "I need to stretch my legs after being in that saddle all day. We could take a walk along the lake."

"Sounds great," Starr replied. "May I ask you one last question? No more after this. Promise."

"Let's go," he answered and walked away.

Starr caught up to him. "How did the people out here survive the bad water and air?"

"Caves," David answered, striding briskly along the shore.

"Caves?" Starr asked, trying to keep up. "What do you mean?"

"Up in the mountains. The air was cleaner. There were underground pools and streams."

"What did they eat?"

"Food!" David said as he stopped and turned around. "Look, I don't have all the answers. No one does. But I can tell you this. People can be very creative when their lives depend on it."

Starr decided to drop the subject; they walked together in silence. It was a beautiful night. The moon laid down a path of light across the surface of the lake and the waves plopped softly along the shoreline.

"David, I'm sorry if I brought back bad memories for you," Starr said contritely.

"Forget it. Maybe it's best to remember to keep the memory of the horror alive so it will never happen again," he said. "Let's hope man has learned from this."

"Where will we go tomorrow?"

"The first village. It's a two hour ride."

They traveled the north shore of the lake and the first thing Starr noticed shortly after the village came into sight was the smell of the pigpens. "What's that awful smell?" she asked, holding a scarf over her face.

"This is the village of the Pigkeepers," David replied. "It supplies the whole region around here with pork."

"How do they stand the smell?"

"Pigs can get used to anything I guess," David replied.

"What do you mean?"

"Wait'll you meet the Pigkeepers. You'll find out."

They rode down the narrow, muddy road between acres of squealing, oinking, snorting pigs, hogs, sows and boars. The pens were alive with hundreds of them, all sizes and colors, lying down or moving about in a sea of mud. The racket was deafening.

A man, pulling a wooden wagon that held several large barrels, walked toward them. He was about Starr's height and weighed about three times as much. His baggy jacket and pants were dun-colored and a wide-brimmed felt hat was pulled down on his forehead.

When they were fifty feet from the man it hit Starr like an invisible wall. Her stomach turned over and she thought she would vomit as she reined her horse. David looked at her with a slight smile on his face, then took a deep breath and rode toward the man. Even the horse began backing away and shook his head as if the stench were tormenting him.

Starr settled the horse down and watched as David spoke with the man at the wagon. Beyond them several children were playing in the streets. They looked as much like misshapen mudballs come to life as children. But they were running and squealing and teasing each other the same as any children would do. The village itself was a double row of identical sod houses lining both sides of the road. The grass of the sod walls had long since decayed and blown away. They were, in fact, mud houses with wooden roofs. Smoke curled from the roofs of most of the houses.

Two hundred yards north of the village was an immense, low-roofed building made of wood. From inside it came the shouts of men and the terrible shrill cries of the hogs. Dozens of black iron pots were smoking on fires in front of a series of large open doors facing to the south. Women and older children were moving about in regular patterns as if they had been doing it all their lives. The women and girls busied themselves around the pots, while boys led small groups of hogs into the building and pulled laden carts out of it toward the rows of pots.

Starr sat on her horse and held the scarf to her face, trying not to think of what was going on inside the building. She saw David take a packet of dried fruit from his knapsack and hand it to the man with the wagon. Then he turned his horse and headed back toward her while the man opened a narrow gate and pulled his wagon into an alleyway that ran between the pens.

"I think I may have found the man we're looking for," David said as he approached.

"David, please," Starr said through her scarf. "Let's get out of here before we discuss anything."

"This isn't so bad," he said.

Starr was astonished at his remark. "Are you demented? What could be worse than this?"

"Our next stop," he answered, pointing toward the north. "The village of the Dung Gatherers."

David saw Starr's eyes widen above the scarf. She turned the palomino and galloped out of the village at full speed. "Hold on!" David yelled, racing after her. "Starr, wait a minute will you?"

He caught up to her at a grove of trees near the edge of the lake where he grabbed the reins of the horse and pulled it to a stop.

"No! No! No!" she shrieked. "Absolutely not! I refuse to go to a place like that."

Starr had closed her eyes and was shaking her head. When David made no remark she stopped and glanced at him. A smile was slowly tugging at his lips.

"I mean it, David!" she said sternly. "I'm not going there."

David was laughing now.

Starr realized what had happened. "You did it to me again, didn't you?" she said furiously.

David's head was thrown back and he was roaring with laughter now. He dismounted with some difficulty and sat under a tree.

"I don't see what's so funny about it," Starr said, getting down and walking over to him.

"You're absolutely right. It's not funny at all," he said with tears streaming down his face. Then he rolled over on his side and broke into another spasm of laughter.

"Well, if you're not going to take this seriously, we might as well go home," Starr snapped, sitting down under the tree. She was surprised at the way she used the word *home.*

David gained control of himself and sat up. "I'm serious now," he said. "I'm truly serious."

Starr looked into David's eyes, still glistening with tears of laughter, and thought how much he looked like a little boy. "I guess it was sort of funny after all wasn't it?" she asked.

David looked back at her with a soft smile. "I didn't mean to make you mad," he said. "It's just that I've been worried that you might get hurt out here. I think I just needed a good laugh to break the tension. Works every time."

Chapter Eleven

A Kiss After Dying

On their way to the Badlands, David and Starr rode through the Sand Dunes north of the River. They had been formed thousands of years before by the winds picking up sand from dry river beds. As the wind was constantly changing the shape of the Dunes, there were no permanent landmarks to get one's bearings. Thus, many travelers wandered there until they died of thirst or exposure.

It was now mid-afternoon and Starr was covered by a fine layer of sand. She had followed David's example and tied a cloth about her face to protect her nose and mouth. She shielded her eyes with her left hand. Starr found herself blinking constantly. With the sun blotted out she had no sense of direction and she could barely see David a few feet in front of her. So, she closed her eyes and trusted her horse to follow David's.

"How much further?" Starr called out.

"It won't be long now. We're only touching

the edge of the Dunes."

Thank God for that, she thought.

Starr closed her eyes again and let her mind drift with the easy motion of the horse. Something pulled her from her reverie and she suddenly realized the horse had stopped. Opening her eyes she quickly scanned the area in front, then on all sides. Nothing!

"David! David!" she cried, but her voice was lost in the storm. There was nothing but the dreadful moaning of the wind and the driving, blinding sand. She urged the horse ahead to try and catch up with him. *Surely he's missed me by now. He'll be coming back any second.* Panic was beginning to overtake her. Her breath was coming in shallow gasps and she couldn't control her thoughts.

Then she saw him! A man in a pale cloak riding an ivory-colored horse. *But he's going the wrong way! Directly toward the interior of the Dunes.* She shouted at him, but he was out of earshot. *I've got to stop him. He'll die in this storm if he keeps going.* She whirled her horse around but could gain no ground on the man in the pale cloak. In a few minutes time he was out of sight.

Starr was desperate now. *I'm too far into the Dunes for David to ever find me! I'll never get out of this place!* An overpowering fear took her and she quietly wept as she gave in to despair.

She was choking and her eyes were filled with grit. Untying the cloth, she wiped her eyes and face.

"I told you not to get separated from me!"

Starr looked up. She felt unreal. *How could this be David?*

"What's the matter with you?" he demanded.

"Oh, David, I thought I was lost forever in this place."

"What are you talking about? I only missed you a few seconds ago. I thought you were right behind me."

"I got lost. I rode the wrong way," Starr said breathlessly. "I followed a man and he led me back to you."

"What man?"

"Didn't you see him? He was riding a light colored horse."

"I didn't see anyone," David said. "The Dunes can do strange things to your mind, Starr."

"He was real, David," she insisted. "If I hadn't followed him, I never would have found you."

"All right. Settle down. We'll talk about it later. Right now it's time to get out of this place."

An hour later they rode out of the Dunes into the Badlands. To their right, a towering sandstone butte ran northwest, disappearing into the purple shadows of the mountains. In the afternoon sun it glowed like a giant ember risen from the fiery depths of the earth. They turned west, riding into the sun through a dry rock-strewn plateau, thus avoiding the narrow twisting canyons that scarred its surface.

"Do you really think this man will know where my parents are?" Starr asked.

"I don't think the Pigkeeper was lying," David replied. "He's not sophisticated enough to learn that particular skill. But his information was old and may not be much help to us now."

"How could they possibly be alive after all these years in this awful place?" Starr asked.

"People are a lot more durable than you think, Starr," David said. And as an afterthought, "Your father must be a good man for them to exile him to this part of the Fields. Either that or he had to use this place as a refuge."

As they rode Starr gazed at the snow-capped peaks of the mountain range to the north. They were gleaming in the sunlight like cold fire. *There is beauty in all this desolation,* she thought.

Toward evening they came to a small gathering of huts. Their walls were made of stone and the roofs were rough-cut timbers sealed with pitch. In front of some of them fires were smoking in pits dug in the ground. Women wearing long drab-colored dresses and scarves or bonnets were cooking in black iron pots. Men in coarse trousers and jackets or slouched together in twos and threes strolled on what passed for a street.

David stopped his horse at the edge of the first hut. "Wait here," he told Starr, handing her his reins. He took the few steps over to the first fire and squatted next to a woman who with a wooden ladle was stirring a mixture of black beans and corn.

Starr heard him ask about the man named Assad. The woman with no teeth looked up from her cooking and smiled at him. "Why you want that little worm?"

"Information," David answered.

The woman turned back to her cooking. David took a silver coin from his jacket and began turning it over in his fingers. It glinted in the firelight and caught the woman's eye. Then she looked away, but it drew her head slowly back like a heavy winch. She pointed at a building on the opposite side of the street. David tossed the coin at her feet. Scooping it quickly into her hand, she dropped it into her apron.

"Stay here with the horses," David said,

unbuckling the leather strap on his knapsack.

"I want to go with you," she insisted.

David reached into his knapsack, took out a wide-bladed knife with a bone handle, and slipped it inside his belt under the jacket. "Someone has to watch the horses."

"But I don't want. . . ."

"Do as you're told!" David barked, pulling his staff from its leather case alongside the saddle.

His words stabbed at her chest. She had seldom been spoken to like this, and never expected it from David.

David saw the pain in her face and took her hand in his, looking directly into her eyes. "I'm sorry," he said softly, "but out here you have to do as I tell you. There may come a time when your life — both our lives will depend on it."

"I understand. I'll do better," she said shyly.

"This shouldn't take long," he explained. "If you hear me call you, bring the horses to the front of that building. If I'm not out by the time you get there — wait."

David turned and walked into the dusty street, slinging the staff over his left shoulder by its strap. The sun was sliding behind a low range of hills, casting an orange glow over the village. As darkness gathered, shadows flickered along the walls of the buildings in the smoky light. As she watched David walking among the fires, Starr remembered Martha's prayer about angels and wished there was something she could do to protect him.

Inside the makeshift saloon it was darker than the twilight David had just left. He stood at the door and let his eyes adjust to the gloom. A bar to the left was nothing more than a rough plank laid across two barrels. An oil lamp sat on it and another hung from the low ceiling. Two men stood at the bar drinking from pewter mugs. Three more sat at a table in the opposite corner playing cards. They did not look like the type who would enjoy family picnics.

David walked to the near end of the bar while keeping the five men in his line of vision.

"What'll it be?" the bartender asked.

"Assad," David replied and saw the eyes of a small man at the table dart toward him.

Assad lived by his wits and missed little that went on about him. He had watched from the door of the saloon as David tossed the silver coin to the woman. He talked about it with the men who sat with him. Their plan was to relieve David of the burden of any additional coins he might be carrying.

The bartender made no reply to David, but

glanced at the table in the corner and then walked to the other end of the bar.

David observed the small man in the dark robe. The sharp angles of his face were shadowed under the hood and his dark eyes had a malignant glint. His companions were "bookends" owned by a giant with a morbid sense of humor. Both had wiry red hair that curled like Medusa's about their heads. Even their eyes had a reddish tint to them. They wore filthy black jackets that matched the color of their teeth perfectly.

"I'm looking for a man and a woman," David said flatly, staring down at the small man. "I was told you might be able to help me."

Assad smiled at the "bookends." "Helping isn't exactly my line of work. Maybe you should find the nearest priest." The three of them thought this exceedingly funny.

When they finished laughing, David produced a gold coin from his jacket and held it between thumb and forefinger, three feet from Assad's face.

"What I know is worth more than that," he responded as his right eye twitched almost imperceptibly.

David sensed it was coming and now he knew when.

Some men are born with that combination

of speed and power that are as rare as the cardinal virtues — David was such a man. Had he been born in OldAge, he could have excelled in any sport — but he was not. He was born in an age where survival, rather than touchdowns and home runs, made timing and quick reflexes essential. What he excelled at was staying alive.

A thin-bladed knife appeared in the hand of the man nearest him. David stepped back with his left foot, the right hand reaching across his body to grip the staff three quarters of the way down. His left arm pulled free of the staff's leather strap as the hand reached behind his back and found the bone handle of the knife.

The thin-bladed knife was aimed at David's heart as the man holding it lunged forward. David uncoiled from his crouched position, pivoting on the balls of his feet as his right shoulder turned with the arm following through, whipping the staff around with a backhand motion. It exploded against the side of the man's head one inch above his right ear, crushing the skull like an eggshell. He dropped like a side of beef. Less than a second had lapsed since he pulled his knife.

David's left arm continued around from behind his back and forward, hand gripping the knife, wrist bent inward. The second man was on his feet, swinging a heavy short-handled axe toward David's chest with his right hand. At that moment, David flicked the knife with a vicious motion of his forearm and wrist, his body continuing around away from the path of the axe. He hit the floor, catching himself on his left hand and spinning to put his back to the wall, the staff held in front of him. Two seconds had lapsed.

Blood was seeping through David's jacket where the axe had grazed his left shoulder. The man who had thrown it stood perfectly erect, eyes bulging as he gripped the handle of David's knife with both hands and tried to pull it from his throat. A gurgling sound came from his open mouth and red froth bubbled down his chin as he collapsed across the table.

Assad was frozen to his chair. He was not the type of man to do his own dirty work. David had counted on that. The two men at the bar returned to their conversation.

David picked up his gold coin from the floor and stood before Assad. "I believe you have some information for me," he said evenly.

Starr almost shouted for joy when she saw David leave the saloon and walk across the street toward her in the last of the light. "Did you find out where my parents are?"

"Assad knew them, but. . . ."

"David, your shoulder!" Starr interrupted him.

"Let's get out of here now," he said, mounting his horse. "We'll fix it later."

As they rode away from the village, David continued. "Assad hasn't seen them in years. He told me there's a man close by who may know something."

"Do you think he was telling the truth?" Starr asked eagerly.

"I believe he was," David answered. "This one time anyway."

They stopped outside the village at a small stream that meandered along the base of a hill. Starr watched as David cleaned and stitched the wound himself. "How can you stand to do that to yourself?" she asked, horrified at the sight.

"No doctors — no hospitals out here. You do what you have to. Everyone carries a first aid kit with them when they travel."

Starr was obviously shaken at what had happened. The violence of the City was usually clandestine, but the rawness of this world dismayed her. "David, what happened back in the village?" Starr asked. "I mean — was anyone else hurt?"

David had finished the stitching and was putting the needle and thin coil of gut away.

"Yes," he said and his face mirrored the anguish in his voice. "Sometimes there's no way to avoid it."

Starr sensed that satisfying her curiosity was not worth what it would cost David to talk about it. "Thank you for helping me look for my parents, David," she said sincerely. "This has been a whole new experience for me out here in the Fields — people helping each other and expecting nothing for it."

David looked at Starr with a quick smile. "That must be some place you live in, that City," he said, swinging into the saddle.

As they rode up the hillside on a winding path, a night breeze was rustling the leaves of the stunted oaks. In the distance an owl was calling. It was time for the hunt to begin so he lifted from his treetop with a silent, deadly power.

"Who is this man who knows my parents?" Starr asked.

"His name is Lazarus. Assad seemed to think he was exiled from the City years ago."

Around a sharp turn of the path they came to the cabin abruptly. It was built like those in the village, only smaller. The single front window was glowing with a dull yellow light. A thin stream of smoke drifted upward from the chimney.

The door opened inward and Lazarus

appeared like something out of a dream. He had a long full beard and shoulder length hair. Both were white, the same color his robe had been years ago when it was new. He was nimbused by the lamplight from behind him and his expression was that of someone listening to ethereal music. Starr thought of sixteenth-century paintings she had seen of men who had been labeled by the church as "saints."

"Welcome to my home," he said with one hand on the door and the other outstretched toward them.

"Do you know him?" Starr asked quietly.

"Never laid eyes on him before."

They sat on leather-backed chairs before the fireplace. Lazarus lay propped up on pillows on his narrow bed against the wall. The only other furniture was a crudely made table and a wooden chest that sat in the opposite corner. The single room was warm and cleaner than Starr had expected.

In the light of the lantern that hung from the ceiling David could see the sallow color of Lazarus' face and the pale eyes that were clouded with pain. "How long before you go home?" David asked him.

Lazarus' eyes took on a far away look, a longing to break free of mortality. "God knows," he replied wearily.

Starr was puzzled. "This isn't your home?"

Lazarus looked at David, who made no response, then back to Starr. "I'm dying, child," he said in a kindly voice. "David means my eternal home."

Again Starr was mystified by the way these people regarded death — like it was no more than opening the front door of a house and going outside. "I'm sorry," she said awkwardly.

"Oh, don't be, child! I've grown weary of my role as stranger and pilgrim. I long to be with my own."

David noticed Starr shifting about uneasily in her chair and asked Lazarus about her parents. Starr relaxed as the conversation changed direction.

"I knew them in the City before we were all banished to the Fields," Lazarus responded, his eyes lost in memory. His face turned to Starr. "Your father was one of the first leaders of the Christian Movement. The City Fathers branded us 'Disciples of Treason.' They spent two years looking for your father. He and your mother were both sent to the Relievers, but at the last minute the City Fathers were afraid this would make martyrs of your parents, so they banished them instead."

"What are these 'Relievers?' " David asked.

Lazarus told him. "It's their ultimate

method of control," he said, finishing his explanation. "Anyone considered a threat can be declared a dissident and executed."

"That's barbaric!" he exclaimed. "I thought they were more civilized than that."

Starr was visibly shaken.

"I'm sorry," David said to her. "I know you can't control what your government does."

Lazarus explained that he had not seen Starr's parents for many years. He believed they were still living in one of the villages in this remote part of the Fields. David and Starr visited with him for a while longer, saying they would make camp outside and leave early in the morning. Lazarus prayed for them to have a safe journey and that Starr would be able to locate her parents.

The fires in the village below were dying out one by one as Starr made beds of pine boughs and leaves. David had built a fire and was preparing their first hot meal since leaving Haven. After eating, David and Starr treated themselves to cups of coffee.

Starr was agonizing over David's reaction to the "Relievers." She had come to believe that it was an honorable profession and that she was making a contribution to her fellow citizens in the City. "Our worlds are so different," she said.

David sipped his coffee thoughtfully. "How

else could it be, between slave and master?"

"Why do you talk like this?" Starr la-

"Because that's how it is. You think we chose to come out here?"

"But you and I haven't been like that. Like slave and master. I won't listen to this anymore," Starr said, resting her head in her hands. "I'm not like that. Let's talk about something else."

David lay back on his blanket, looking up at the night sky. Starr fought against the emotions that were tearing at her. How could she remain loyal to the City and feel so strongly attracted to this "Primitive" and his way of life. She looked at him lying next to her and the words were spoken as if by someone else. "David?"

His eyes were closed. "Hmm," he said, stretching his lean frame and turning his back to her.

"David, is there anyone —" she paused. "Anyone special in your life?"

"Dozens," he said without turning around.

"I'm serious, David."

He turned over, sat up and rested his arms on his bent knees. "What's this all about?" he asked.

"I'm — I'm interested in the customs of your people, relationships between men and women, courtship patterns. It's for the research I'm doing."

"You first," David said with a tomcat smile.

"We don't have courtship."

"What do you have?"

"If two people consent, they sign a 'Loving Friends Contract,' " Starr said with a trace of arrogance.

"What in the world is that?"

"They live together for a specified period of time, not to exceed six months. Under certain conditions, it can be extended if they both agree."

David was appalled and it showed on his face. "That's barba. . . . That's very interesting," he replied calmly and lay back down.

"Well?" Starr asked indignantly.

"Well, what?" David teased her.

"What are your customs?"

David got up and walked to the brow of the hill. "Ah yes, our customs. I'm afraid they're painfully boring next to yours."

Starr walked over to him. The moon had risen, casting its pale light on the winding stream below. Starr thought it looked like a silver necklace shining at the base of the hill. "I'd like to hear about them," she said. "For my research."

David looked down at her upturned face in the moonlight. "If a man and woman have affection for one another, courtship can follow. The man begins by. . . .”

“Only the man can initiate it?” Starr interrupted.

“I'm afraid so. It's his choice.” David paused thoughtfully. “At least we like to think it is.”

“What's next?” Starr asked a little too eagerly.

“He asks the father's permission to see her. Then he'll bring her flowers or a gift. They go for walks, have picnics, go to family gatherings, sit in the parlor and talk — any number of things.”

Starr was thinking of flowers — flowers and sitting in a parlor with David.

“Starr, are you listening to me?”

“Yes, yes, go ahead.”

“Well, if things work, courtship can lead to marriage and a family.”

“How long are these marriage contracts for?”

“Life.”

Surprisingly this didn't shock Starr, although she knew logically it should have. She had the warmest feeling in her breast — warm and yet somehow cold, for she was beginning to tremble.

“Are you all right?” David asked, taking both her shoulders in his hands.

Starr felt weak. She moved close to David and pressed against him, holding his waist. His arms went around her with a gentle strength. Starr was frightened at the feelings welling up inside her. She tilted her head back and traced the scar on his face with her finger. She felt his hands caressing her back, her neck and touching her face. She was on tiptoe now and his lips were touching hers with a warmth that moved like a current slowly through her body and returned to the place where they were joined as one person, one breath, one life.

"No!" something seemed to scream deep inside Starr. This can't be. Not with this — Primitive. "I shouldn't have done that. Forgive me," she said and turned away.

"Of course."

That moment would return to them again and again for the rest of their lives. The time they stood together on the hilltop in that wild land with the moonlight streaming through the trees and their lives were indelibly changed.

Chapter Twelve

A New Life

At dawn a rain began to fall, causing Starr to awaken with a start. The sudden miniature thunder of the fat drops on the roof pulled her out of a sound sleep, and for one instant panic claimed her. Wildly she looked across the room half expecting to find Miriam awake, but the dark-haired girl was sleeping peacefully. Then she sat up, staring at the rain falling in thin silvery lines. The drops that struck the window ran crazily down, some meeting others in an abrupt joining.

Starr lay back, listening to the drumming on the roof close to her head. It was the first time she'd lain and listened to the rain, and as was true with so many experiences she'd had since coming to the Fields, she was filled with a faint sense of regret. The opaque Dome that enclosed the City had served to protect the population from the deadly air and water produced by the Greenhouse Effect years ago — but now it screened the things she had found most beautiful. Open skies, running brooks, the rumpled soil, and the activity of bird and beast had become a delight to her. Now the sight of the rain falling aslant and the patter of drops on the roof brought pleasure to her. *I'll miss it all!* she thought, then suddenly felt that she was betraying her way of life.

Closing her eyes, Starr settled herself under the blanket. Soon the sound of the rain brought sleep again. When she awoke, Miriam was gone. Starr could hear the faint sounds from below — the women speaking quietly, the noise of dishes and pans floating up the stairs. Throwing back the cover she arose, washed her face and dressed. Then she took the VoiceWriter from her case, but hesitated slightly before she began her report. After she'd gathered her thoughts she pressed the switch and began.

"Starr Omega. Report Number 26 to Remedial History Section. Attention Emmett Tau. Subject — Mystic and Romantic Tendencies of Primitives. This report will give instances of actual behavior patterns resulting from those mystical and romantic strains delineated in Report Number 23. Basically, this element (which seems to be inherent in some of the subjects) is most clearly seen in two areas — religion and courtship. The religious activities will be dealt with in Report Number 27, while this report will document the 'romantic' element that makes up the character of the primitive psyche."

Starr pushed the *Pause* button, then walked over to the window. For a few moments she stood there watching the rain as she organized her thoughts. Then she activated the Voice-Writer and said in a flat voice, "The mystical strain that evidences itself in courtship patterns is not easy to isolate, though the results are evident. The primitives believe in something which they call 'falling in love.' This phenomena is almost unknown in the modern world, but its roots go back to ancient times. Basically 'falling in love' is antithetical to the reasonable and logical processes which exist among the civilized today. On the surface the phenomena can be observed when a male and a female begin to turn from others and find pleasure in each other's company. In the City this occurs, of course, but enlightened people base this on personal need, and are not bound by any commitment other than a Loving Friends agreement. Basically, in the City the individual entering into this sort of contract is saying, 'I need you and will do whatever I have to do to get you.' Hard as it is to believe (and even more difficult to understand!) the Primitives who have 'fallen in love' are saying, 'I want *you* to have something.' Obviously, this cannot be true, for all modern scientific study by our sociologists indicate that human behavior is totally self-centered, and much of our educational processes are devoted to teaching individuals to get *what they want and need* from others, no matter what must be done to the other individual."

Here Starr paused momentarily, thinking of a couple she had met on a field trip. David had taken her to a small cottage seven or eight miles from the village, saying just before they got off their horses, "I think you'll find this couple interesting." Luke, the husband, was tall, strong and about twenty years old. He had married a girl, a year his junior, named Dorene. Starr had heard that Dorene was one of the most beautiful girls in the village and sought after by many young men. But she had chosen Luke, and they had married. Only six months after their wedding, their small house caught fire, and though Luke had escaped with only minor burns, Dorene had been terribly scarred. David had taken Starr to their home, and the sight of Dorene's twisted, ravaged face and claw-like hands had repelled her. Starr saw that she was expecting a child very soon, and had kept the revulsion from her expression by an effort. She had been shocked when the ruddy handsome Luke had put his arm around his wife's waist and kissed her scarred cheek. When they had left the cottage, Starr asked, "How can he bear it, David? She's hideous!" He had given her a strange look, then said, "Love is not love which alters when it alteration finds." He said no more than that, but Starr guessed his meaning. "You mean," she had asked with a puzzled frown, "that love doesn't change when something bad happens to the other person?" He nodded, but the whole affair haunted Starr. She considered putting the story into her report, but was certain that her superiors would never believe it.

For half an hour Starr spoke clearly and concisely, letting no emotion register in her voice. A memory flashed through her mind as she told how a young man with a bouquet of flowers had come to call on Miriam. She recalled how she had run across such things before in her research, and was aware that something in the simple action appealed to her. Quickly she said, "It is my theory at this point that such behavior has its roots in a certain body of primitive literature which has somehow survived in the Fields. This literature consists of poetry, especially that of a medieval poet called Shakespeare, and stories called 'Novels' most of which date back to the last period of OldAge. Such material has long been eradicated from the City — except for those copies which are peddled by pornographers — but they do exist in the Fields.

"One of the most popular of these I located in the small collection of the family I am staying with. There was no cover, so the title and the author are lost. The story concerns a man with the strange name of *Rhett* and a young woman with the equally bizarre name of *Scarlett.* The action takes place in the distant pre-industrial past of OldWorld, and is almost completely devoid of logic. What it does contain is the strange magnetism or attraction that Scarlett possesses. Young males are drawn to her like moths to a candle flame; they apparently possess no more sense! She treats all her *suitors,* as they are called, like dirt; which doesn't seem to bother them at all. They keep coming back to her, unable to resist the powerful force that she exudes. The character called Rhett is different from the other males — and in some ways is just as attractive to females as is Scarlett to males. It is a long book, concerned with some civil struggle, but basically deals with the phenomenon that centers on the relationship of Rhett and Scarlett. She is obviously a very silly young girl, and the book never explains how a man as powerful and handsome as Rhett is willing to put up with her — this is where *falling in love* finds its expression. Logically, they should not become involved. But they cannot seem to help themselves."

Starr paused again, then continued: "From my study of these 'romantic' works, it seems to me that 'falling in love' is somewhat like a virus. The subject in this condition, when in the presence of his or her lover, is described as being short of breath, flushed in the face, stricken by a heart that beats abnormally fast and hands and limbs that tremble unaccountably. The classic symptoms of a virus! However, those who 'fall in love' react as though this experience is the epitome of human existence, for when they are separated from their lover, they go into a psychological and emotional pit of despair. This despair is quite illogical, for there is no shortage of available mates. Why these primitive people, on losing a partner, cannot simply move on to another is a subject that defies the logical mind. End of report. Starr Omega."

She pushed the button abruptly when she heard Miriam's voice calling her down to breakfast. Stowing the VoiceWriter back into her case, Starr left the room to take her place at the table beside Timmy. "Hello, Starr," he said with a bright smile. "I wish my hair was curly like yours."

Starr laughed and ran her hand through his straight black hair. "Maybe I'll give you a permanent, Timmy," she said.

"A permanent what?" he asked in surprise.

"Oh, that's something that makes your hair curly."

"Do *you* have one?" he demanded.

"Oh, no. Mine just grows this way. But I think we'd better not make your hair curly. I think it looks nice the way it is."

"All right," he answered agreeably. Then he picked up a biscuit and bit off a mouthful. "Will you read to me some more today?"

"I will if David and I get back in time. He's going to take me to see some people."

"Can I go, too?"

David had been sitting back watching the two. "Not this time, Timmy, but tomorrow after the service, Starr will take you down to the pond if it isn't raining."

"Oh, that'll be good!" He launched out on a rambling tale of how he and David had caught a turtle in the pond; and the grown-ups made no attempt to interrupt. Once again Starr was struck by the obvious love the other members of the family had for Timmy. She wondered, but not for the first time, if they realized that he would not be with them for long. *They must know. Surely there have been other Down's Syndrome children born here.* But she said nothing, and after breakfast David said, "Let's get on our way, Starr."

"Don't be late, David," his mother said.

"You don't want to tire Starr out."

"Oh, David is quite entertaining on the buggy rides we take," Starr said innocently, but with a gleam in her grey eyes. "He always —"

"Time to go!" David broke in hurriedly, and taking Starr's arm he almost pulled her out of the room. "We have lots of chores to do."

When they were in the wagon, she asked impishly, "I didn't have a chance to tell your mother about how you behave on buggy rides."

"Never mind that!" he said hastily. "I told you, Mother is pretty old fashioned. Why, she's told me lots of times that she never kissed any man but my father — and him only *after* they were married!"

"Oh? I wonder why you fell into such bad habits?" Starr teased him lightly as they drove along the muddy road, enjoying the sheepish look on his face. Finally she asked, "Where are we going today?"

"As I told you, we have quite a few members of the Ecclesia who need visiting. Some of them are widows, some are elderly or sick. The men of the Ecclesia visit them pretty often. See that they have firewood, fix anything that's broken — things like that."

Starr remained silent for a time, then said

softly, "I think that's nice."

"Nice? Well, I guess so. Never thought of it that way."

It was a fine day, the sun coming out to warm the earth, sending its beams over the pools standing in the fields. The road was muddy, and the hooves of the horses splattered the red mud over the buggy, including some that came to adorn Starr's face. David observed that she looked better with it, and then let her have his big handkerchief to wipe it off.

They spent all morning visiting, most of the stops were at the homes of elderly people. Everyone greeted David with warmth, and all were curious about Starr. With the frank curiosity of old age several of them asked, "Is this your sweetheart, David?" At first Starr was nonplussed and felt her cheeks redden, but David laughed at her. "No, this is my old aunt here on a visit," he would say. Once he said, "No, she's looking for a husband, but she told me right out that I wasn't handsome enough for her." The shrunken old lady he said this to had stared at Starr with displeasure, saying pertly, "Well, you're just too choosy, Missy! This boy is fine looking enough for you or anybody else."

"See?" David had said as they were back in the wagon. "You're too choosy. I'm fine looking enough for any young woman."

"Too bad *you* think so!" Starr retorted, but glancing at him, she thought, *Well, he is good looking! And he doesn't know it, not really.*

A little after noon they stopped and David pulled a basket out of the back of the buggy. "Nice place for lunch over there," he said. Hopping down he reached up his hand. At first Starr thought he wanted her to hand him something — then realized that he was waiting to help her down. No one did such things as that in the City. Women were proud of their independence and would have taken such a gesture as an insult. But as she put her hand in his she discovered that the act made her feel — womanly was the only word for it. "Thank you, Rhett," she smiled up at him.

"Who?"

"Oh, just a name," she said quickly. "Let's eat over there." She ran to a large tree that had fallen, and soon the two were sitting on it eating their lunch. As always, she tore into the food as if she were starving. David grinned at her as she chewed and swallowed the thick beef sandwiches, popping hard-boiled eggs into her mouth like peanuts. "Want mine, too?"

"Oh, this is so good!" she exclaimed, ignoring his teasing. "What's in the little wrappers?"

"Fried pies."

Starr opened one of the packages and her mouth watered as a delicious smell came to her. "I thought you *baked* pies."

"Sometimes you do," he answered. "Sometimes you fry them in grease." Watching her taste the pie, then begin to eat it with huge enjoyment, David said, "I better eat mine or you'll take it away from me." He watched her eat two of the pies, then added, "I don't see how you stay so trim. As much as you eat you should be as fat as old Jezebel." Jezebel was the huge sow that came grunting to the fence for handouts. David gave her a close inspection, then shook his head. "No. You've got a much better figure than Jezebel. At least, that's what Jonah Logan told me."

"He didn't say that!"

Jonah Logan was a young friend of David's who had come to the house several times lately to visit. He was shy, but Starr knew that he liked her. "Not exactly. He said you were very pretty and asked if he could come calling on you." David carefully ate the last of his fried pie, then dusted his hands regretfully. "I told him he couldn't."

Starr's head jerked around to face him. "You told him *what?*"

Ignoring her indignant outburst, he said calmly, "I told him you went around kissing men in buggies. He's a fine lad, and I don't want him to pick up your City ways."

Starr's jaw dropped, and she began sputtering. "Why, you — !"

David shook his head. "Jonah's going to marry Beulah Wright. She's been in love with him since they were ten years old."

"Well, is he in love with *her?*"

"Not yet, but he will be."

"Oh? He will be? And how do you know that, may I ask?"

David began stuffing the remains of the lunch into the basket. When he was finished, he looked at her and said evenly, "God gave me assurance that he would. Come on, we've got a lot of visits to make."

His reply silenced Starr for some time. Only after they had made three more visits, and he announced that they had only one more stop, did she bring up the subject. "You say *God* told you that Jonah and Beulah would get married? What did He sound like — God I mean?"

David said easily, "Starr, God doesn't speak to us with a voice. He's a spirit. He can use any method He wants, but with me it's like an *inner* voice. And usually when God speaks to me, it's something I've been thinking and praying about. For example, I'd been worried about Jonah and Beulah. He's a little flighty, you know. Apt to be impressed by flashy women." She glared at him, but he ignored her. "So I began praying for them, and after a time I began to get this assurance. Can't really explain it, but I'd been in doubt — and then for no reason, I just began to *know* that Jonah would wake up to what a fine girl Beulah is."

"Maybe you just *thought* that yourself, David."

"I doubt it. Nothing to make me think it. Jonah hasn't changed. He's still running around after young women. Still doesn't give Beulah a thought. No reason why I *should* know he's going to marry her. But he is."

Starr asked after a long silence, "Did you ever feel that God had spoken to you — and find out you were wrong?"

"Sure."

His ready assent caught her off guard. "Didn't that — make you *doubt?*"

"A little, I guess. But the longer I serve Jesus, the better I know what He wants." He seemed to be struggling with words, and suddenly turned to her, his dark eyes intent. "Starr, the longer you're close to somebody, the better you know them. Isn't that so?"

Starr could only stare at him. Finally she whispered, "David, I don't know. I — I've never had anyone that I could really come close to."

David was shocked by her reply. He kept his eyes fixed on her, then shook his head. "That's not right," he said gently. "You've got to let people get close to you. We all have our little boundaries, Starr. Walls we build because we're afraid that if people get close they'll hurt us."

"Well, won't they?" she asked sharply.

"Some will, sure. But I'd rather get hurt than be a hermit and live alone."

His answer struck her hard, and he saw her lips tighten. "God never meant for us to be alone, Starr," he said gently. "He sets the solitary in families."

Starr had no answer for that, but changed the subject. After they had driven for two hours she asked, pointing across a pasture, "Isn't that where the girl who was burned lives?"

"Luke and Dorene," he said. "There's Luke —" He broke off abruptly; for the man had started running toward them. "Something's wrong," he said, and slapped the reins on the horses. Coming even with the man, he jerked the horses to an abrupt stop. "What's wrong, Luke?"

"It's the baby, David!" Luke cried. "It's coming early."

"I'll go for my mother!"

"No time!" Luke said, grasping David's

arm. "She's afraid, David — and I guess I am, too!"

David leaped out of the wagon. "It's God's baby, Luke. I've heard Dorene say that many times. Well, God will help bring the child into the world. Better get back to her." Luke wheeled and ran to the house, and David said, "Come on, Starr!"

Looking down at the hand which David held to her, Starr was filled with apprehension. "I — I'll wait in the buggy."

But David reached up and lifted her out of the buggy as if she weighed no more than Sarah's kitten, Bobo. "Time to break down one of your little walls!" he said roughly.

"But, I don't know anything about birth!" she practically squalled, and even as he pulled her across the yard, a line from the book about Scarlett and Rhett almost spoke itself in her mind:

Miss Scarlett! I don't know nothing about birthin' no babies!

Suddenly he whirled and grasped her by the shoulders. His eyes burned into hers, and his voice grated, "You're a woman, aren't you? Or maybe not. I've been wondering a lot about you, Starr. Now I guess I'm going to find out. Are you just a machine taking notes and spitting them out? Or are you a real live woman?"

"That's not fair!"

"It never is! What do you want to do? Get in the buggy and ride off? Leave them alone?"

Starr tried to break free, but his grip was like steel. She pleaded then, "David, I can't help! We need a doctor!"

"Dorene needs love and someone to hold her. Someone to tell her she'll be all right. You can do that." Then seeing her expression, he suddenly loosed her. She almost fell, but he said, "No, I guess you can't. You don't know what love is. And you never will!"

His words cut into her, and anger welled up at once. Who was he to tell her she didn't know love! She whirled and ran away from the house. A small grove of trees bordered the cabin, and she moved toward it her pulse throbbing. But the silence of the tall trees did not stop the voice that kept ringing in her ears: *You don't know what love is. And you never will!*

She had entered the grove at a fast clip, anger raging through her; but it slowly ebbed, leaving her with a barren emptiness. The shadows of the towering trees made a gloomy sanctuary and the silence of the grove was broken only by a mournful cry of a bird, faint and far away. Finally she slowed her pace, then came a strange sense of loss. It was bitterness, she recognized at once, and then she understood that David's words had touched a raw spot that had been in her all the time; dormant and hidden even from her own consciousness. The sense of poignancy that came to her was like the plaintive cry of a violin she had once heard, a sound that had brought tears to her eyes, though she never understood why. She knew that David had spoken the truth.

"I *don't* know what love is!"

She started, for she had spoken aloud, and the sound of her voice in the silent glade sounded loud. Despair came to her and as she looked up, her eyes filled with burning tears, she cried out, "How could I know love?"

No answer came, and Starr stood there as the echoes of her voice faded. She remained motionless for so long that a small furry animal she'd never seen before came out of the underbrush, sniffing eagerly and grabbing green shoots. He came right up to her feet and almost touched her — then his entire body seemed to become electrified! He threw himself backward in a paroxysm of fear and tore across the vine-covered earth, scrambling with a desperate intensity to escape the Other Thing that had invaded his world.

The fear of the animal touched Starr and she whispered, "You're afraid, too, aren't you?" Then she lifted her head and the muscles of her jaw grew tense. Slowly she nodded, then turned and made her way back to the clearing. Without a pause Starr walked up to the cabin. The sound of a woman's voice crying in pain came to her. Starr hesitated for only a moment — then pushed the door open and entered.

Both men were standing beside the bed, and they both turned to stare at her. The woman's eyes were closed and her scarred lips were sealed as she pressed them together. Slowly she opened her eyes and looked at the newcomer. Fear was there, but more than that, a plea for help for someone to share the pain.

Starr moved across the floor, then dropped to her knees beside the bed. Taking the frightened girl's hand, she reached out with her other and pushed a strand of hair back from the pale forehead.

"You mustn't be afraid, Dorene," she said quietly. "You'll be all right — and so will your baby!"

She spoke with a firm assurance that she didn't feel. But Dorene's hand clamped down on her own, and hope appeared in the pain-filled eyes. "I can't help it!" she whispered.

"We'll help you," Starr said gently, and even as she knelt there assuring the woman, her mind was busily reconstructing the pages of a book she had once read. A book that dealt with primitive medical techniques. And Chapter Seven had been entitled, "Childbirth."

Starr looked up and said, "Get some water boiling — and all the clean towels and cloths you can find — !"

By the time David pulled the horses to a stop, the sun was breaking over the trees to the east. He silently admired the line of crimson light that traced the contours of the eastern hills, then turned to look down at Starr.

She was lying in the crook of his arm, her face pressed against his chest. "Starr — we're home," he said. Her long eyelashes fluttered, and then her eyes opened.

"It's morning!" she exclaimed. Then she realized that she was lying in the curve of his arm and quickly pulled away. Her limbs were stiff, and she stretched to relieve them. The memory of what had happened came rushing back to her. He watched her mobile lips move slightly into a smile. "It was like nothing I've ever seen, David," she murmured, her eyes bright with the memory.

"You did fine, Starr."

"Oh, I didn't do much," she added quickly.

"I didn't want Dorene to know it, but I was more frightened than she was!"

"You sure didn't show it," he remarked. "A beautiful little girl."

Starr thought of the moment when she had cleaned the baby, noting with wonder the perfect fingernails, the completeness of the small body. She had never seen such a thing, and it had given her a sudden sense of despair to recognize that she had "relieved" many mothers of babies not much younger than the one she held in her arms.

And when she had placed the child in Dorene's arms, the sight of the smile on the ruined face took away all the ugliness of the scars. The miracle of birth had shaken her thinking about the meaning of life. She knew as she sat there with David, that no matter what else, she would never again be able to abort a child so long as she lived.

"Guess this is where I apologize." She looked up with a startled expression, finding him watching her with a serious look. "About what I said — that you never knew what love was," he explained. "I was wrong about that."

Starr could not speak for a moment, then she said slowly, "No, you were right, David. I don't know what love is."

"It's what brought you into the cabin to help Dorene when everything in you fought against it," he answered. "You didn't do it for money or recognition. You came to help Dorene in any way you could. And to tell the truth, if you think *you* were scared, you should know what Luke and I were feeling! We were absolutely paralyzed! Starr, if you hadn't been there I hate to think what would have happened!"

She merely shook her head. But as he helped her down from the buggy he asked, "Do you know what Dorene told Luke she wanted the baby named?"

"No. What was it?"

He held her for a moment as she came to ground. Then as she looked up at him, waiting for his answer, his broad lips curved in a pleased smile. "She named the baby *Starr* — which I thought was appropriate." Then he said gently, "Another little Starr for this world. Makes me feel very good. Come on, let's tell the family about it."

She smiled briefly, but her mind was filled with the image of the tiny baby resting on the bosom of the young woman with the scarred face — and of Luke, the strong, handsome one, looking at both of them with love and pride in his eyes.

Chapter Thirteen

An Unwelcome Message

Exhausted physically and emotionally after the crisis with the birth of Dorene's baby, Starr wanted desperately to fall into bed and rest. But this proved to be impossible, for she and David arrived just as the household was beginning to stir. Miriam and her mother were finishing the cooking, and soon the room was humming with the sound of talk around the breakfast table. David and Starr were kept busy answering questions about the new baby and the parents. The young couple were good friends of Caleb and his family.

David minimized his own efforts by saying, "It's a good thing Starr was there! Luke and I were useless, but Starr knew just what to do."

Starr was uncomfortable with the admiration she saw on the faces around the table. "Oh, Dorene is the one who did all the hard things," she said quickly. Starr had finished her eggs and bacon, but when she said, "I think I'll rest for awhile," Timmy protested.

"You said you'd go to the Meeting with me today! Aw, you can sleep anytime."

Starr started to shake off his pleas, but there was such disappointment on his face that she hadn't the heart. "I guess I can sleep later," she surrendered, and went at once to shower and change clothes. Miriam came in just as she was getting dressed, and taking a look at her said, "We usually wear dresses to the Meeting, Starr. I know you don't have one, but you can wear this one if you like. We're about the same size."

Starr took the dress at once, not wanting to violate any local code. Slipping into it, she took a quick look in the mirror and was pleased at what she saw. The dress was a simple gown, light blue with a finely wrought collar of white lace. It clung to her figure, and she said, "What a nice dress, Miriam!"

"Oh, it's just one I made myself. It looks nice on you, Starr." She herself was wearing a light brown frock with a full skirt that swept her ankles. As she brushed her hair, she spoke about the new baby, and it seemed to Starr that she was somewhat envious. "I can't wait until I get married and have children, can you, Starr?" She ran the brush down her thick mane of glossy black hair, not noticing that Starr did not respond. "I'm going to be late starting my family. Most of my girl friends are already married." Putting down the brush and getting to her feet Miriam smiled ruefully, "I'll be married this year, though."

"Will it be Saul?" Starr asked.

"Oh, no!" The answer came quickly, and Miriam shook her head emphatically. "I'm not in love with him. Besides, he doesn't want a wife — he wants a mother for his two girls. And there are lots of women around who'd welcome the chance to marry him. Several widows would jump at the chance."

"Who will it be, Miriam?"

Miriam smiled at Starr, then said with a slight smile, "I'll marry Nathan."

"Has he asked you?"

"Not yet, but he will when I give him the sign."

Starr looked up with interest. "The sign? What's that?"

"Oh, you know, Starr!"

"No, I really don't."

Miriam gave Starr a strange look then said, "It's not any one thing. But when a girl wants to let a man know she approves of him, there are ways she can do it."

"Why not just say it?" Starr asked seriously.

"Oh, no! You have to be more subtle than that," Miriam protested. "You have to become very attentive to him. Smile at him more, and once in awhile when he helps you down from the buggy, you hold his hand a *little* longer than necessary. . . ."

Starr listened with astonishment as Miriam described the patterns of courtship, and finally when the girl was finished, Starr shook her head, getting to her feet with a short laugh. "It's very complicated, isn't it?"

Miriam suddenly bit her lip. "No, not really. I'm making it sound like it. But when a woman loves a man he'll be able to see it. How could he not?" Then she shook her head and laughed, "Well, the lessons on how to deal with men will have to continue when we get back. Come on, Starr. Mother hates to be late for the Meeting!"

David pulled the buggy to a stop beside a long hitching rail. "That's where we have our Meetings," he said, nodding toward a single-story structure made of peeled logs. "We're a little late this morning."

He helped Starr down, then reached up and swung Timmy to the ground. "I want to hear you sing nice and loud this morning," he said putting his hand on the boy's shoulder. Then David led the way to the front of the rectangular building, followed closely by the rest of the family who had come in the wagon.

As Starr entered the building she looked around curiously. She saw a single room broken on both sides by three windows that admitted bright rays of light. The long seats were made of planed lumber, and most of them were already filled with people. “There are three seats down front,” David whispered. Starr followed him reluctantly, for she would have preferred a less conspicuous place at the back of the room. She noticed the looks she received as they walked to the front and sat down. She tried to ignore them.

As soon as they were seated, Timmy whispered, “It’s about time for the singing, Starr. That’s my favorite part of the Meeting.”

Even as he spoke, one of the men got up and stepped up on a low platform at the front. He said nothing, but lifted his voice and began singing in a clear tenor voice. At once he was joined by the entire congregation and the sound filled the room. Starr did not know the song and wondered if the group met to practice. Finally, she determined that they learned the songs by repetition. She sat there listening carefully to the words, looking down once in awhile at Timmy, whose face was rapt as he sang in a piping voice. The words were:

Make a joyful noise unto the Lord, all ye lands.
Serve the Lord with gladness:
Come before his presence with singing.

Know ye that the Lord he is God:
It is He that hath made us, and not we
ourselves.
We are his people, and the sheep of his
pasture.
Enter into his gates with thanksgiving and
into his courts with praise:
Be thankful unto him, and bless his name.
For the Lord is good; his mercy is
everlasting;
And his truth endureth to all generations.

The song was sung several times, and soon Starr found that she could sing along with the others. She had a quick ear and a fine contralto voice. Timmy looked up at her, his eyes warm, whispering, "You sing *good,* Starr!"

The song service continued for an hour. There were no solos, which seemed strange to Starr, and the musical accompaniment consisted solely of a man and a woman who were on one side of the room playing guitars.

Perhaps it was the very *simplicity* of the music that touched Starr; she was accustomed in the City to highly complex and complicated forms of music. Here the music seemed an extension of the people who stood and lifted their songs with faces radiant and eyes filled with obvious joy. Most of them were poor people, wearing colorful but plain clothing. Starr had seen how hard they worked and how little they possessed. Why were they so joyful? A joy that filled the room in the form of song.

Her professors had taught her that religious emotion was simply a psychological response. *Learn how to push the right button in people's psyche* they had said, *and they'll respond. Just like Pavlov's dogs who drooled when a bell was rung.*

But something in Starr denied this, since there was no stimulus to cause these people to behave as they did. Some of them raised their hands in a simple gesture of longing, some clapped their hands when the song was fast, and many of them had faces stained with tears — tears of happiness rather than of sorrow.

As the service continued, Starr struggled to understand it — until something happened that was past *understanding*. The leader started a new song, the words simpler than the ones he had already sung:

> Jesus! Jesus! Jesus!
> How I praise the name of Jesus!
> Fairer than the Morning Star,
> He's everything to me!
> Jesus! Jesus! Jesus!
> He fills my deepest need,
> Only Jesus!

For no reason that she could comprehend, as the song went on, the word *Jesus* began to have a strange and disturbing effect on Starr. The first indication she had was when her voice began to falter as she sang the song. Then, deep within her a poignant sorrow began to make itself known, and she could say that name no longer. All her life she had trained herself to keep any deeper emotions under careful control, and even as the sadness and voiceless grief grew stronger, she set her lips and kept her face stiffly fixed.

But she could not control one thing — the tears that rose unbidden to her eyes. Almost angrily she willed them away, but the singing continued softly, and every time the name Jesus sounded, it was like a small dagger into her heart. Fiercely she bit her lips to keep them from trembling, but the tears would not be controlled. To her horror she felt the hot drops running down her cheeks, and quickly she dashed them away, hoping no one had noticed.

But her hope was unfounded, for David had caught the motion, and without a word he took a handkerchief from his pocket and handed it to her. She took it and removed the tears, then tried to think of something else — anything to keep from breaking down in public!

Finally the song dwindled and fell away, and without a sign from anyone that she noticed, the people all sat down. Gratefully she sank into her seat, took a deep breath and determined that the emotion that had swept her would never be mentioned in an official report. She could imagine what a cold fish like Bernard Alpha would make of such a thing in one of his researchers!

She was fairly well under control when a man got up and faced the congregation. He was not impressive, being only slightly above medium height and having a plain round face. But there was something in his light blue eyes that caught her, a hint of some sort of inner fire that she had seen a few times — mostly in poets and artists.

But his voice! She had never heard such a voice in her life! It was not loud, but she knew that if he had chosen to lift it above the almost conversational level he used, it would have filled the small building like an enormous bell!

And yet — it was not the sort of voice used by professional politicians which overpower the hearer. No, it was really an intimate voice, warm and informal. The sort of voice one would stop reading to listen to. Without meaning to, Starr found herself listening to the man, and soon she was drawn into what he was actually saying. He began by saying, "The portion of the Book that I will read from this morning may seem strange to you, but it has been a comfort to believers for centuries. . . ." Then he explained that what he would read was part of a vision that God had given to a man named John after the death and resurrection of Jesus. He began reading from a book, and though the language was almost mystical, she found herself caught up in the imagery of it:

And I saw in the right hand of him that sat on the throne a book written within and on the backside, sealed with seven seals.
And I saw a strong angel proclaiming with a loud voice, Who is worthy to open the book, and to loose the seals thereof?
And no man in heaven, nor in earth, neither under the earth, was able to open the book, neither to look thereon.
And I wept much, because no man was found worthy to open and to read the book, neither to look thereon.

He paused, then looking out with compassion in his direct blue eyes, said, "He wept. Why? Because he wanted someone to help him — but he knew that no human being was able to help. That is the problem of all of us, even of kings and emperors. We long for something beyond what we have, but search as we may, we never find it. We search the philosophers and the thinkers of the past, and find that they did not have the answer, not even for themselves. Many try to avoid the question, throwing themselves into pleasure, but they still weep inwardly. . . ."

As he continued to speak Starr had the frightening sensation that he was speaking directly to her! His eyes fell on her from time to time, and it took all her will to keep her gaze steady. *He knows!* she thought desperately. *He knows all about me!* The horrifying fear came that he would point his finger at her and lay bare all that she had managed to keep hidden from the eyes of others.

But he did no such thing. After he had spoken of the search that men and women made to find peace — which always failed — he said in a stronger voice, "But there is one who can open the book! Every book, even the book of your life which you've kept hidden from everyone! For the rest of the passage says:

> *And one of the elders saith unto me, Weep not: behold, the Lion of the tribe of Juda, the Root of David hath prevailed to open the book, and to loose the seven seals thereof.*
>
> *And I beheld, and, lo, in the midst of the*

throne, and of the four beasts, and in the midst of the elders, stood a Lamb as it had been slain, having seven horns and seven eyes, which are the seven spirits of God sent forth into all the earth.

And he came and took the book out of the right hand of him that sat upon the throne.

And when he had taken the book, the four beasts and the four and twenty elders fell down before the lamb, having every one of them harps, and golden vials full of odours, which are the prayers of the saints.

And they sang a new song, saying, Thou art worthy to take the book, and to open the seals thereof: for thou wast slain, and hast been redeemed up to God by thy blood out of every kindred, and tongue, and people, and nation;

And hast made us unto our God kings and priests: and we shall reign on the earth.

Then the preacher lifted his hand toward heaven and raising his voice like a trumpet, he shouted, "Who was this one who alone could open the book? Jesus! Jesus Christ, the son of God!"

The congregation began to cry out, "Jesus!" and what Starr had felt before was nothing to the sudden surge of emotion that erupted in her! It was fear and joy and hope all at the same time, and she sat there, trembling in every nerve, knowing that she could no more stand to her feet than she could fly out of the window!

The preacher began to speak, and she heard him as if she were a long distance away. What she did hear was the name of Jesus, many times, and each time that name was spoken it was as if a sword pierced her heart. She did not understand the words that had been read, and only faintly did she grasp that the preacher was saying that this person, Jesus Christ, was God, and that somehow He wanted something from every man and woman.

She felt Timmy's hand touch hers, and grasped it blindly, not daring to look at him. She felt his other hand come to rest on hers, and somehow that gave her a great comfort.

Finally the voice ceased to speak, and she was able to stand to her feet with the others. Dimly she heard someone speaking a prayer of some kind. Then David was leading her outside. She was aware that a few people spoke to her, but she could never remember what she answered. She was able to come up with a smile and some sort of response, but it was a relief when she was alone with Timmy and David as they drove slowly back down the road toward the house.

Timmy chattered happily as the horses plodded along through the muddy track.

David spoke from time to time, but Starr heard little of it. She was still shaken by the storm that had swept over her, leaving her drained and weak.

Finally they came into view of the house, and she looked up when David said, "I think you've got company, Starr."

Looking up quickly she saw a tall man lounging under a tree, his great stallion tied to a branch. When they got closer she said in surprise, "I know him. He's one of the Border Guards."

"Will Sigma," David nodded. "One of the better ones."

The guard waited until they all got out of the buggy, then pulled a piece of paper out of his tunic pocket. "Message from the City," he said without expression. His shaggy blond hair and eyes the color of flint made him look tough, but he stepped back after delivering the note.

Starr's hands trembled slightly as she fumbled with the envelope. A sense of foreboding seized her, and she saw that the note consisted of one line and a signature. When she read it, a sense of fear came over her, for it said what she had most dreaded to hear:

You will leave at once with the bearer of this message and report to me. Sheila Phi.

Chapter Fourteen

To See Jesus

Looking out the window of Miriam's room, Starr was packing her saddlebags which were laid across the bed. She thought of the peaceful nights of drug-free sleep and of rising to a bright morning refreshed and eager for the day to begin. The times around the breakfast table with the family as they shared the delicious home-cooked meals before going their separate ways for the day, the lowing of the cattle coming across the evening meadows, and the sound of music and laughter in this home were things she could not bear to think of losing.

Below, Martha was preparing coffee and Starr could hear the conversation of David and Will Sigma as they sat around the table. It still surprised Starr that this family could treat even a Border Guard the same as they would an old friend who had come for a visit. As she finished packing and descended the narrow stairs, Timmy came running to her.

"Starr! Starr!" he said breathlessly. "You

told me you'd take me to the pond if it wasn't raining. Remember?"

"I remember, Timmy," Starr said, rumpling his hair. "But something's come up and I have to go back to the City now." Starr looked at Will who was shifting uneasily in his chair.

Martha brought the coffee service to the table and began pouring the rich, steaming liquid into their cups. "I'm sure Mr. Sigma has time to eat a piece of my blackberry cobbler," she said, looking at Will. "It'll be ready in a few minutes."

"I can't think of anything short of a firing squad that could keep me from it," Will said, smiling broadly. He looked at Timmy standing with his arm around Starr's waist, then nodded to Starr.

"Thank you, Will," Starr said as she left the room with Timmy holding tightly to her hand. "We won't be long."

The afternoon sun was pouring into the valley, filling it with pale gold light as Starr and Timmy walked hand in hand across the pasture toward the pond. Spindly-legged calves were frolicking and wobbling among the scattered trees, never straying far from their mothers. From the south came the distant dry rumble of thunderheads forming and Starr knew they would bring rain by evening. She had come to love the rain, to love its whisperings among the leaves of the trees and its cool breath through her window.

She remembered the first time she had seen the painting that hung in her apartment and believed that a world like this could never exist. She had found that world and now she was losing it. Timmy's hand was warm in hers, but what she felt was a stone cold aching in her breast. The land around her was green and growing and full of life — she saw only the stark barren towers of the City. *Oh, God, I can't go back there! To that dead place! I can't leave all this!*

"What's wrong, Starr?"

Timmy was pulling on her hand and Starr knew he had sensed her distress. "Nothing, Timmy. What could be wrong when I've got a friend like you?"

Timmy beamed up at her. "Watch what I can do," he said excitedly, collecting a few small stones from around the water's edge. Starr sat on a rough wooden bench under an ancient gum tree and watched while he tossed them into the pond. Near the far shore trout were feeding under overhanging willow branches, dimpling the surface of the water.

In her mind Starr was back in the City. She saw Philemon — Martha Epsilon as her head turned slowly to the side and she fell asleep, her face framed by the ragged white hair and faded blue shawl, beautiful in death — and others! Memory battered her heart until the pain flowed upward from her breast and coursed down her cheeks in hot rivulets. A massive silence engulfed her and she found herself gasping for breath. All she had valued, all she had longed for had come to nothing but loss and grief.

"Starr."

She heard Timmy's voice as if from a great distance, so great she could not reach him.

"Starr," Timmy called, walking over to her. "Why are you crying, Starr?"

Starr put both hands to her face and began wiping the tears away, unable to speak.

"I know what will make you happy," Timmy said softly. "I'll sing a song for you. It's the first song I ever learned."

Starr took a deep breath and managed to control her sobs. "I'd love to hear it, Timmy," she breathed.

"Mama taught it to me when I was real little."

Starr smiled and took both of Timmy's small hands in hers as he began to sing in a clear voice. The song of a Meadowlark came to her across the pond, a breeze rustled the leaves above them and set the water shimmering in the sunlight — as an old woman,

Starr would remember every detail of the next moments with perfect clarity.

Jesus loves me this I know.

Starr looked into the face of this child who would be despised and rejected in her world and she saw the Wisdom of the Ages.

For the Bible tells me so.

In the weakness of the child she saw a power greater than any she knew existed.

Little ones to Him belong.

She saw peace — perfect and eternal.

They are weak, but he is strong.

Starr closed her eyes and a silent cry rose from the depths of her soul. Oh God, help me! I'm sorry! Have mercy on me!

Yes, Jesus loves me.

She sat there listening to the simple song, and when Timmy finished, he looked up into her face. "Starr, do you know what?"

"No, what?"

"When I was real little, I cried, just like you."

"Did you, Timmy?"

"Sure!" His eyes were open and trusting, and he was very serious. "I wanted to have Jesus in my heart — like Poppa and Momma and David."

Starr asked, "What did you do, Timmy?"

"I asked Jesus to come into my heart!"

She waited, but he obviously felt he had

given her the answer. She hesitated, then asked, "What did you say?"

"I said, 'Jesus, I don't like the way I am. I want you to make me happy, like Momma and Poppa. So come on into my heart.' " Beaming up at her, he nodded with absolute assurance. "That's what you've got to do, Starr."

"But, Timmy, that's too easy!"

"Sure it's easy! If it was *hard,* I couldn't have done it!"

His words came to Starr like a blow! She suddenly realized that she was hearing the absolute truth about Jesus. She had assumed from what she had heard at the Ecclesia and from the members, that what they did was do the best they could. But that had never rung true to her. She knew full well that her best was not good enough for God. And she knew as well that her life had been very wrong, filled with things that were totally contrary to all that she had seen and heard in the Fields. No matter that they were accepted in the City — she *knew* there was such a thing as right and wrong!

Her problem had been one of understanding. Over and over she had asked herself the question *How can a man who died centuries ago bring anything to me?* But she knew from her observation of the members of the

Ecclesia that those who followed Jesus Christ were different. But *how* to make it work — that had been her downfall.

She sat there beside Timmy, everything in her saying no! to what was happening. *If I did become a follower of Jesus, I'd lose everything!* she thought. And she flinched from the thought of being an outcast. *But how can I go back? How can I face life in the City, when the only real happiness I've ever known has been in the time I've been here? And how can just asking Jesus to come into my heart make any difference?*

Her mind whirled — yet deep within her spirit something was pulling at her. She felt very certain that the decision she faced was not one which she could reverse, and it frightened her — and yet — it frightened her worse to think that she might pass by the only reality she'd ever known.

"Come on, Starr," Timmy urged, pulling at her hand. "It ain't hard. Just tell Jesus to come in and fix you up."

Starr knew that it was her "moment of truth," as the old seers had called it. And with a sense of desperation, she closed her eyes, and feeling very foolish said huskily, "Jesus, I ask you to come into my heart. Cleanse me from all that's dark and wrong — and make me as innocent as this child!"

She opened her eyes — and saw the face of

Jesus! How could she know this Jesus — see him in the face of a child? How could His love flow through her like warm oil — healing every wound — bearing every sorrow — cleansing all sin? How could there be such a pure love, accepting her completely in spite of all she had done, demanding nothing in return? She knew she would never be alone again behind her wall.

Starr remembered as a child when her father would toss her high into the air and she would fall, it seemed forever, before he caught her. Now she let go of her life completely and fell into the arms of One who would hold her forever. She was never able to explain or even understand what happened to her that day. In all the years to come, she would speak of it as "The day I saw Jesus."

Timmy finished the song and looked at the tears streaming down Starr's face. "You're all right now, Starr," he laughed. "Those are happy tears."

Starr remembered the words David had spoken to her that morning on the bluff. *The wise are confounded and the children understand.* "You're right, Timmy. Your song has made me very happy."

"That's because it's about Jesus," Timmy said, clapping his hands. "He's my best friend."

David was saddling Starr's horse when he saw Starr and Timmy walking up from the pond to the house. Martha was waiting for them on the back porch. She spoke with Starr for a moment, then hugged her goodbye. Timmy did the same and Martha took him inside the house. As Starr walked along the path from the house to the barn, David could tell there was something different about her. When she took his hand and looked up into his face, he knew what had happened.

"I don't know how it happened, but you truly are part of our family now, Starr Omega," he said, taking both her hands in his and smiling warmly down at her.

"I'm so happy, David! The world seems different now."

"It's because you've been changed, Starr — by the presence of Jesus Christ living in you. The emotion you feel may fade, but His love never will. This is truly the beginning of your life. You need to remember, though, others will be angered at this change in you. It happens to every Christian. But we endure through the strength that we have in Jesus. We can't do it ourselves."

"Oh David, how can I leave you! How can I ever live without you!"

David drew her close to him as she circled his waist with her arms and pressed her face to his chest. "There can be no goodbyes for us, Starr," he said, brushing her soft hair with his lips. "We'll be together again. But for now, you must go."

"I can't, David! I can't!"

David touched her chin with his fingertips and turned her face up toward him. "I'll come to you, Starr. I promise."

"But how?"

"I don't know yet. But I will," he assured her. "For now remember — you have a friend who will never leave you or forsake you."

Starr became quiet. She was confused and afraid, yet in the deepest part of her being there was strength and a sense of peace she had never felt before.

"We can't always understand why things happen to us, Starr," David continued. "But the Bible tells us 'that all things work together for good to them that love God, to them who are the called according to his purpose.' Now let's go. Will's waiting for us."

"Think of me, David," Starr said as she touched his face and looked into his dark blue eyes.

"You're part of me, Starr. Part of my life." He kissed both her cheeks softly, then her lips. "Being apart won't ever change that."

Starr had no words for what she felt. She couldn't bear the thought of leaving David and the life she had found here. The City and all she had known there seemed like a monstrous perversion of what life should be. There was a dreamlike quality about it and she began to wonder if it actually existed.

"Sorry it had to happen this way," Will said, sitting astride his great stallion. "Nothing I could do about it though. I'm just a messenger boy."

"I understand," Starr replied. David held her hand and she bent to kiss him, then rode away with Will across the pasture among the grazing cattle. She looked back once, and as her eyes met David's she felt her heart lift within her. At that moment she knew that she would always be a part of him, that he always would be with her. And Starr felt the love of God flowing within her like a river of pure light, driving out the darkness where fear and doubt and sorrow dwell — and she remembered, as she would do for the rest of all her life, opening her eyes and seeing Jesus in the face of a child.

"Do you know why they're demanding my return?" Starr asked, with a note of concern in her voice. "I'm barely halfway through my research. It doesn't make sense."

They were riding together toward a distant valley with the sun falling toward the purple hills in the west. Their long shadows slipped silently across the contour of the land ahead of them like dark serpents drawing them back to the City.

"They never tell Border Guards the why of anything, little girl," Will answered, moving easily in his saddle with the gait of the horse. "We're little more than slaves to the City Fathers."

Starr was amazed that anyone would speak this way openly. "Aren't you afraid they'll find out how you feel? Those words are tantamount to treason."

"How could they find out?" Will replied, looking all about him at the open countryside. "No one here to tell them."

Starr smiled at him. "I guess you're right," she said.

"We're all of a kind, little girl. Puppets clop-clopping about at the whim of the almighty stringpullers," Will said, throwing his head back and jerking his arms around. "Get people to swallow the big lie, and the string- pullers can control millions."

"You mean that people have to stay inside the Dome to be safe?" Starr asked.

"Exactly. And drugs are the final touch. Make them accessible and no one's got the gumption to stand up for anything," Will said with a growing anger he was trying to control.

"Some of us think there are Domes like ours scattered all around the country. Every one with Manuals and Primitives to do the labor."

"Will," Starr asked, still shocked at his outburst, "if you feel this way, why don't you just leave? You do have a certain amount of freedom to come and go."

"You *are* naive, Starr Omega," he said, shaking his head slowly. "A lot of us would like to, but our families are little more than hostages back there. Oh, they treat us well enough — but they make it clear what would happen if someone deserts. They even keep us in our own separate community so we don't infect the general population with what we know about the world outside the Dome."

"It is nice out here, isn't it?" Starr said, trying to steer Will away from the troubling turn their conversation had taken.

His face brightened. "Nice, little girl?" he said, looking about him at the rolling green hills. "Why it's the Garden of Eden."

"Where did you hear about that?" Starr asked, surprised again at Will's words. "Oh, I pick things up here and there," he said, winking at her. "Come on, I'll race you to the creek."

They camped by the creek that night and the sound of it rushing over the stones was the earth whispering to the star-filled heavens.

The cry of a whippoorwill calling its mate carried on the evening breeze.

Starr sat with her back against one of the pines in the grove where they had made camp. She thought of what she would face when she returned to the City. All her life she had managed to avoid most of the bureaucratic pitfalls, keeping her career on track. It was the driving force in her life. Now it all seemed unimportant, insignificant, and she desired only to be free of the City and return to the Fields. She thought of Martha Epsilon's wish to go home to her "beautiful green Fields" to die.

Starr looked at Will thoughtfully sipping his coffee while the firelight played on his sunburned face. She had washed their supper dishes and put them away, to his surprise. "My, my," he had remarked. "The city girl's learned something worthwhile."

"Will," Starr began, "why can't I stay out here? Couldn't you tell them you just couldn't find me?"

Will laughed softly, eyes intent on the fire. Then he looked at Starr. "You really have led a sheltered life, little girl."

"Well, why not?" Starr demanded.

Will's face became troubled. "Starr, I saw how things are between you and David. If there's anyone who could keep you safe out here, it's him. Why do you think he didn't try to stop you from going back?"

"Well, you've got that shotgun."

Will laughed again. "David could make me eat this shotgun if he wanted to."

"Why didn't he do something then?"

"Because he didn't want you to die, or any of his family or friends. And that's exactly what would have happened."

Starr looked at Will wide-eyed.

"If you and I aren't back by a specified time, a helicopter gunship will be dispatched from one of the firebases just outside the laser wall and Haven would be nothing but ashes and smoke the next day." Will let his words sink in, then continued. "They did that a lot in the early days of the Fields. Hardly ever happens now. The Primitives meet their grain quotas and the City pretty much leaves them alone."

"No one ever said anything about this," Starr reasoned, "About what could happen to them."

"You grow up 'under the gun;' you don't think about it. You just accept it as a fact of life."

Before she drifted off to sleep Starr saw clearly that her job as a Reliever was as heinous as anything the Border Guards did with their gunships — *but that part of her life was gone forever!* In her dreams, she stood once again with David on the hilltop outside Lazarus' cabin as the moonlight drifted like spun silver over that wild and frightful land, then she smiled in her sleep. In feathered silence, an owl soared above her, its talons bloodied from a night's work.

The metal barge shuddered, then jerked forward to the sound of creaking pulleys as it moved out into the River. Starr and Will stood next to their horses and watched the dark water breathe a white vapor into the night air. On the far shore Starr could barely make out the squat figure of Sheila Phi standing in the yellow swatch of light next to the guard station at the ferry landing.

"Welcome home, Omega," Sheila Phi said as Starr led the palomino down the ramp to the bank. "My, don't you look positively radiant after your sojurn among the Primitives?"

"Thank you," Starr said brusquely. "Why was I recalled?"

"All things in due time, Omega," Sheila Phi replied, standing with her legs apart and her arms folded over her ample bosom. "For now, you just come with me."

"What about my horse?"

"Leave him," Sheila Phi said sharply.

"You'll have no need of a horse from now on."

As Starr was untying her saddlebags, she noticed Sheila Phi motion quickly with her hand and a Peacemaker seemed to materialize in the darkness near the guard station. He walked over and grabbed the saddlebags before Starr could react. Throwing them over his shoulder, he headed toward a buggy drawn by two black horses.

"Come, come, Omega," Sheila Phi murmured in a husky voice. "We'll have a nice talk on the way back." She took Starr by the arm and led her over to the buggy.

The only sounds were the clopping of the horses' hooves and the creaking of the buggy springs as they rocked through the night toward the distant gleam of the City. Starr was lost in thought when she felt Sheila Phi's hard stubby fingers stroking her shoulders.

"You must be exhausted after your trip, dear," Sheila Phi said, breathing heavily.

Starr tensed at her touch, but remained silent.

"You'll feel much better when you're back in civilization among your own kind."

Sheila Phi's fingers were stroking Starr's neck and when her hand slipped down to unbutton the blouse, Starr shuddered and pulled away with a gasp.

"You dare to reject me, you little trollop!" Sheila Phi snapped, sitting upright in her seat. "I think some time in the Tank might persuade you to be a little more cooperative!"

Oh God! Starr thought. *What have I done now!* She fought back the tears that were welling up inside her.

Sheila Phi took a deep breath and regained her control. "Don't think your reports have escaped our scrutiny, Omega," she said severely. "I felt it my duty to bring it to the attention of the higher authorities."

"I don't know what you mean," Starr said quietly.

"I think you do. Let's just say for now there are indications that you have become — shall we say, particularly enamoured of this spurious philosophy that's rampant among certain groups of the Primitives."

Starr remained silent.

"Christianity, Omega," Sheila Phi blurted. "You are familiar with the term, are you not?"

When Starr failed to respond, Sheila Phi crossed her arms over her chest and they rode in silence.

As they neared the City, Starr saw the Dome shining like a great jewel against the night sky. *"It is beautiful from here,"* she thought. Suddenly, as if someone had spoken them to her, the words seemed to ring out in the night, ". . . whited sepulchres, which indeed appear beautiful outward, but are within full of dead men's bones, and of all uncleanness."

Part Three:

THE FLOCK

Chapter Fifteen

The White Tower

The white-suited driver turned the buggy away from the main gate of the City, following a rutted track that ran next to the smooth black base of the Dome. Fifty yards down he stopped before the seemingly blank wall, got out and inserted a thin metal object into a slot. A ten-foot section of the wall slid upward with a slight hum, stopping at a height of eight feet above the hard-packed ground. The buggy creaked from his ponderous weight as the driver climbed back in and drove them into the towering gloom of the Dome.

They traveled along the left perimeter of the Camp where a twelve-foot high sheet metal fence had been constructed to keep the Border Guards alienated from the rest of populace. As they clattered by next to the gleaming fence, a lean man with crew-cut blonde hair, wearing the coarse jacket and heavy boots of the Guards, leaned in a doorway marked Service Entrance. Talking to another guard, he glanced at the buggy with the two women in the back. Starr caught his gun metal eyes with her own and something passed between them that hadn't been there when he put Starr through her training at the Camp.

Although she had lived here all her life, Starr felt like a traveler from another world as they journeyed along this secluded track toward the center of the City. And she sensed something that had been hidden from her before — an almost palpable malevolence that seemed to hover in the dark air. In a few minutes they pulled over next to an open and lighted door set in a dull black tower. Above the door, the single word VERBOTEN gleamed with cryptic intent. The word was used only in direct association with the White Council.

Two Peacemakers stood at attention flanking the sign. In front of them, his bald head shining in the dim light, stood Bernard Alpha. "I see the prodigal has returned. Welcome home, Omega."

Sheila Phi climbed awkwardly down from the buggy, trudging over to Alpha. "Mission accomplished," she beamed.

Alpha held her in the flinty stare of his pale blue eyes. "Excellent, Phi. You managed to stay erect in a buggy seat all by yourself for at least an hour. With progress like that you'll be tying your own shoes in no time at all."

Sheila Phi's face reddened as she shifted about uneasily in Bernard Alpha's less than benign presence.

Starr gazed at the two of them from the buggy, showing no intention of climbing out. *Things are worse than I thought. I've never known Bernard to act this cruelly before.*

Alpha waved Starr toward him. "You may alight from your carriage, Cinderella. The ball is over."

The three of them entered the vacuum tube, flanked by the Peacemakers. They sat in seats attached to the walls, closed the sliding bars around themselves and zipped upward one hundred feet to the Dome Sled platform. They stepped into the sled itself under another VERBOTEN sign.

Starr leaned slightly out of the window as the Dome Sled rose ponderously on its track following the long, slow curve of the roof. Starlight and the brighter light of the full moon shone through the thick acryllic panels in a muted lustre. The blank walls of the buildings and the narrow canyons between them stretched endlessly below her in angular swatches of shadow and light. There was little movement in these early morning hours. A single carrier, diminished by the height of the

Dome to the size of a toy, slid silently along the streets in the vast darkened stillness of the City.

Starr had always had an amiable working relationship with Bernard Alpha and decided to test its waters now. She turned to him quickly. "Where am I being taken?"

Alpha gave her a bleak smile. "To the end of the line."

The White Tower! "Why?" Starr asked, her voice betraying her with its unsteady timbre.

Alpha stared ahead into the murky distance. "You'll find out soon enough."

Sheila Phi remained silent, squirming uneasily in her seat. The two Peacemakers sat like automatons in the rear.

When they reached the final station, Alpha, Phi and Starr disembarked onto the platform. The Peacemakers remained in the sled while it pivoted on its center axis and, with a lurch, began its long descent toward the Rim.

Like most citizens of the City, Starr had never been to the White Tower — nor had she wanted to go. The western station where they stood was one of only four that allowed access, its track extending over to the tower itself — all the other tracks ended from all around the Dome at this circular hub of convergence. Alpha placed his palm on a white circle on the station wall and a narrow door slid open, revealing a tiny four-passenger sled that would carry them over to the tower. "After you, ladies," he said evenly, ushering them into the sled.

As they made the short trip across to the White Tower, Starr looked down more than a quarter of a mile to the walled and gleaming black forecourt that led to its base. Ahead of her two Peacemakers sat in a thick-paneled acryllic guard booth built into the side of the tower. When they were admitted, after the rigid security checks, they walked along a smoothly-polished black corridor to a room somewhere in the interior of the building.

Entering a narrow blinding-white room, Starr noticed a slim man of average height staring at her from a raised, white podium at the far end. His eyes, dark as a ferret's, seemed to soak up the light, and held Starr's with a chilling intensity. As they walked past a gray plastic table with a molded headrest, toward the ferret-eyed man, Alpha motioned for Sheila Phi to remain behind. He then ushered Starr into a slim stainless steel chair directly in front of the podium.

"This is Starr Omega," Alpha said in a level tone.

"You're far more attractive than I was led to believe," Richard Xi said in his precise, clipped speech, glancing at Sheila Phi who shifted nervously about where Alpha had left her. "You look the absolute picture of health. I've never seen anyone other than a Border Guard with a natural tan."

Starr sat erect on the unforgiving steel, trying to avoid contact with Xi's licentious stare. It occurred to her for the first time how pale everyone else was.

"How was your sojourn in the Fields? I understand you made quite an assortment of friends."

Starr cleared her throat nervously. "It was very — enlightening. Vastly different from our society. They have so very few creature comforts."

Xi pressed his hands together in front of his face, rubbing the tips of his joined forefingers back and forth on his pursed lips. "I'd like to hear more about the people — these Primitives."

What's all this leading up to? Why doesn't he just come out and say it? "They lead quiet, simple lives — agrarian mostly, there is some hunting. Everything is centered around the family — the traditional family of OldAge."

Xi gave Starr a bloodless smile. "These — Primitives — what beliefs do they adhere to. They must have some sort of philosophical or religious inclination."

Starr felt a coldness in the pit of her

stomach. She knew full well what happened to citizens under the Dome who had even a nodding acquaintance with the precepts of the Christians. "Some do. Mostly antiquated and disorganized."

"I see," Xi mused. "Well, have you gleaned sufficient information from your foray among the Primitives to complete your dissertation?"

"Yes sir. I believe so," Starr murmured, glancing at Sheila Phi over her left shoulder.

Xi smiled again. "You needn't concern yourself with the approval of your *former* academic advisor, Omega. She'll be going on to — other endeavors."

Sheila Phi took two awkward steps toward Xi, her mouth open to speak.

"That's far enough!" Xi barked.

Sheila Phi stopped like she had bumped into an invisible wall. "But, sir! I've —"

"Silence!" Xi said flatly. "You've proven yourself unfit for any furthur academic service — in fact for any service at all, that I can think of." A cold edge came into his voice. "Your last task for the City was to accompany Omega back from the Fields so her recall would look like a routine student-advisor matter to the Border Guards, in case your accusations against her were correct."

Sheila Phi's face appeared as if it would break apart. Her jaw dropped, the skin on her cheeks sagging like melted candle wax as the icy fingers of horror brushed her. “But I’ve always done as I was told! I informed you of my suspicions!”

Xi fixed his flat dark eyes on her. “Correct! You also, in your inflated egoist manner, told *others.*”

Mystified at the sudden turn of events, Starr revealed it in her expression.

Turning to Starr, Xi explained, “Your stalwart advisor here, took it upon herself to assume some sort of clandestine fact gathering mission regarding your activities in the Fields. She has concluded that, while in residence with a Primitive family, you succumbed to this rather laughable myth of the Crossbearers.”

Starr felt coldness in the pit of her stomach spreading outward until she trembled involuntarily.

Sheila Phi cleared her throat to speak, but was cut off immediately by a sharp glance from Xi.

“Bernard Alpha,” Xi continued, “your superior at the Department of Adjustment has assured me that this could not happen — before your trip to the Fields was approved.” He stared appraisingly at Starr. “I’m inclined to agree with him.”

“Thank you, sir,” Alpha offered.

Xi ignored him. "The facts are, however, that the late Sheila Phi has convinced a small, but highly visible group that such is the case — knowing full well that any information regarding dissident factions is handled by the White Council and is, consequently, subject to the strictest rules of confidentiality."

Sheila Phi stared straight ahead, unable to accept what was about to happen to her.

"Therefore," Xi went on, "you must deny any belief in this alleged Messiah by means of a public forum yet to be decided upon. We can't have the general populace thinking that this antiquated philosophy has any merit whatsoever."

Starr sat with her head down, staring at a crescent-shaped scar in the gleaming floor.

"Bernard tells me you would be most eager to return to work, Omega," Xi stated flatly.

Starr looked up into Xi's pallid face. His eyes held an almost saurian appearance.

"We'll begin with a very small task that could reap some very large benefits for you, Omega — a guaranteed Ph.D. in Remedial History to begin with."

Xi nodded toward Bernard Alpha, who led Sheila Phi to the table. She walked with a leaden gait, her eyes glazed with shock, as she lay back onto the hard surface while her arms and legs were fastened with the plastic straps.

Bernard Alpha laid a white box the size of Sheila Phi's clenched fist on the table next to her. "There is only an endless recycling of the soul into body after reincarnated body," he said to her in a tone that lacked all conviction. In his mind he lay under a willow tree on the cool green grass beside the stream, as he left the room.

Xi rose to follow him and, as he passed Starr he stopped, whispering in her ear. "It's a tiny investment for the return you'll receive from it, Omega."

When the door clicked behind Xi, Starr quickly released the straps that bound Sheila Phi to the table. She helped her to sit up, grasping her by the shoulders and shaking her gently. "Sheila — listen to me! They're going to murder both of us!"

Coming out of her dazed condition, Sheila Phi gazed at Starr with a look of panic in her wide eyes. "No — this is impossible! It's not my time — I'm not through with this life yet!"

"Sheila — you must understand! This is your only life — unless you accept Jesus Christ as your Savior." Starr was frantic, but tried to remain outwardly calm. "There is only one God and he gave his Son, Jesus to die for us. His word tells us: 'For God so loved the world, that he gave his only begotten Son, that whosoever believeth in him should not perish, but have everlasting life.' "

As Starr looked into Sheila Phi's eyes, she saw all the grief and suffering and hopelessness that mankind was heir to. She tried desperately to reach into that dark void. Starr took Sheila Phi's face in both her hands. "Sheila — Jesus loves you so much and he understands your fear and your confusion. Just open your heart to him — trust him — call on his name. Jesus! Jesus! Jesus! Remember to call on that name, Sheila. He understands and he loves you so very much."

At that moment, the door swung silently open. Richard Xi, flanked by two Peacemakers, entered the room. Starr turned from Sheila Phi, looking into a face that was outwardly composed, but in those depthless eyes she saw such an intense hatred that she could almost feel it burning her skin.

"A touching scene, Omega — and a costly one for you," Xi declared bluntly.

As Xi flicked a finger at Sheila Phi, one of the Peacemakers lifted her bodily from the table, escorting her from the room. The other Peacemaker took up a position next to the door.

"I'm afraid we've reached an impasse." Xi paced slowly back and forth, his hands clasped behind his back. "You've obviously been enticed by this insidious and demeaning philosophy. This won't do — this won't do at all."

Starr knew what was in store if she didn't denounce her Christian faith publicly. She remembered one woman in particular who had been sent to her from the Pyramid by way of the Intensive Care Unit. (The Healers always tried to restore their patient's health before they sent them to the Relievers.) The woman had told her: "I was nowhere — I was nothing anymore. There was nothing left for me to hold onto. I saw my soul leave my body. It scuttled sideways like a crab out of my breast and down my arm. It brushed my fingertips — I couldn't hold on — and it fell into the blackness."

In her mind, Starr could still see the woman. According to her file, she was twenty-three years old — she looked seventy-three. Her hair was completely white, her skin hung in loose folds on her bones, and she trembled constantly. But her eyes were the worst. Printed indelibly in them were the horrors that filled her shattered mind. Imagine yourself staring through thick glass into the eyes of a man strapped into a chair bolted to the floor of a tiny room — imagine his eyes as the first smothering wave of cyanide hits him.

"I find it incumbent on me to offer you a

final reprieve before I turn you over to the Healers, Omega." Xi stopped his pacing and held out his hand toward Starr in supplication.

Starr knew what the words would be before she heard them. They were the same ones she had spoken to Philemon and others in what seemed another life.

Xi stood with his hand outstretched toward Starr. "You need only denounce this insidious philosophy that has seduced you and the City will forgive your rebellion, embrace and nurture you."

Starr could still see the assurance shining in the pale blue eyes of the man called Philemon. *Jesus, keep that same light shining in me — give me that same assurance.*

In the Rim's Southeast Sector — Unit forty-three, a thin woman with pale blue eyes, in her mid-twenties sat in tube forty-nine watching the steam begin to rise from the spout of a copper kettle that sat on the small one-burner stove. Her straight, blonde hair was tied at the back of her neck with a piece of twine. She wore charcoal gray coveralls and a threadbare blue shawl, faded from countless washings, was draped over her narrow shoulders.

A rumpled little man with keen brown eyes and close-cropped light brown hair sat across from her, leaning back against the steel curve of the tube. His features were sharp and his voice sounded like it belonged to a boy of ten. "I always think of Martha Epsilon when I see you use this kettle, Philea."

"You know what I think of, Lido?" Philea sighed, her voice soft and fragile sounding. "I think of the green Fields she used to talk about all the time. I dream of them at night and I see the tall trees and the streams and white clouds in a high blue sky. And the bright sun, Lido, and the warmth of it. Do you think it really looks like that?"

Lido smiled, his bright eyes twinkling in the perpetual gloom of the Rim. "Today I saw someone who is just back from the Fields. They forced her to return."

"She went there — out beyond the City somewhere?"

"Verily, verily," Lido laughed.

"Who is she?" Lido's smile ended abruptly. "We've got to help her somehow. I don't know what we'll do."

"Don't let your mind wander, Lido," Philea implored, taking his hand. "Who is she?"

Lido's eyes filled with wonder. "Starr Omega."

"This can't be!" Philea was astounded. "Why would she need our help? Why, she's —"

"She's one of us," Lido declared abruptly.

"But — but how do you know this?"

Lido's mind stopped its wandering. "Because she wouldn't deny Jesus — in spite of what they're going to do to her."

"The Healers?"

Lido nodded.

"Where was she?"

"I saw her in the White Tower."

Philea's eyes grew wide. "How did you reach the White Tower? No one goes there!"

"I go everywhere!" Lido said expansively, spreading his tiny arms. "There's hardly wallspace or crawlspace anywhere I can't go. I go places no one else knows of."

"You're an adventurer, little Lido," Philea smiled. "Most of us just stay to the buildings where we're assigned."

"I can go where Starr Omega is," Lido assured Philea. "I have to help her! She's one of us now."

"She must have heard the Gospel somewhere in the Fields," Philea mused, pouring steaming tea into two blue plastic cups. "I don't think it's gone beyond a few areas of the Fringe here in the City."

"Maybe there's hope for us in the Fields," Lido piped, his eyes shining.

"If we can get to Starr Omega — free her from the White Tower. She could lead us to the Fields." Philea shook her head dejectedly. "No it's impossible!"

"I can do all things through Christ which strengtheneth me." Lido gazed serenely into Philea's eyes, then broke into a smile. "He keeps my mind straight most of the time!"

Philea smiled, her blue eyes filling with confidence. "Thank you for reminding me of that, Lido. It's the only way I could have made it — losing father and Martha on the same day."

Lido sipped his tea thoughtfully. "I'm going to go see Michael Kappa. He'll help me get Starr away from the White Tower. Then we'll have to find a way out of the Dome. He hasn't told me yet, but I think he knows about a secret door."

David stood with his arm around Martha, watching Starr ride alongside Will Sigma across the pasture among the grazing cattle. When they dropped out of sight over a rise, he reached in his jacket pocket for the slip of paper Will had handed to him under the table as they ate.

David,

I'm terribly sorry about this. But I must tell you that if I fail to return Starr to the City a gunship will wipe out your entire village. Also the safety of my family is at stake. In these times we do what we have to.

Starr is in terrible danger. She may be taken to the White Tower and forced to deny her faith publicly. If she refuses — they will break her. They break *everyone* sooner or later.

If you come for her — perhaps I should say *when* you come for her — do as follows:

Wait one full day before you leave in case a watcher has been dispatched.

Leave under cover of darkness.

On the third night, I will wait at the barge landing across from the City.

Call my name once from the edge of the light.

Perhaps some day things will be different. Perhaps someday your King will come.

Will —

David handed the note to Martha. She read it quickly, concern growing in her brown eyes for what she knew her son would do. She hugged David quickly, then stepped back, looking up into his dark blue eyes. "I'll start getting some things ready for your journey. You get all the rest you can until tomorrow night."

David smiled at his mother, noticing the shine of her braided hair as she turned, walking quickly to the house. Then he strode toward the big barn located behind the house near a path that led through the woods to the River. He entered through the main door, turning left into the tack room. Opening a heavy wooden cabinet on the wall, he took out several items and placed them on a table under the single window.

Picking up a short crossbow designed for use with one hand, David slipped the cross-piece into it and strung it to make sure it was in working order. Then he dismantled it, put the stock, crosspiece, string and five metal tipped bolts into a leather case, tying it securely. Next he slipped his oaken staff from its case, running his fingertips over the polished wood looking for any sign of a crack. After checking the leather strap, he put it away in its case.

David closed his hand around the bone handle of his knife, its wide blade glinting in the sunlight that streamed through the window. Sitting on a high stool next to the table, he stroked it gently with a whet rock until the blade was sharpened to perfection. Out of the corner of his eye he spotted a cock-roach crawling in the shadows along the opposite wall. With a flick of his wrist, the knife flashed across the room, quivering slightly in the heavy pine board as it impaled the insect through the center of its back.

The next evening, having told his sister and brothers farewell inside the house where the family prayed for his safety, David stood with Caleb and Martha next to his horse outside the stables.

The distant cry of a whippoorwill carried to them on a breeze across the evening fields where the last slanting rays of sunlight touched the long grass with an amber gleaming. Cattle were lowing as they headed in for the barns. Smoke drifted upward from the suppertime chimneys of Haven, while mothers gathered their children from play as a hen would gather her brood under her wings.

Caleb threw his son's saddlebags over the horse, tying them securely with the leather straps. "Son, where you're going, you'll face dangers far greater than the wolves of the timberlands or even men like Abbadon. In the City you'll enter into a spiritual warfare much deadlier than the physical one."

David looked gravely into his father's face. "I can already feel it, father. It's like a dark wall is being thrown up in front of me — to keep me away."

"We'll be in constant prayer for you until you return, son," Martha promised.

Smiling at his mother, David leaned over and kissed her on the cheek, embracing her. "Then I'll be under a mantle of light that the darkness can't penetrate."

As the sun dropped behind the shadowed hills in the west, David quickly embraced his father, mounted his horse and rode across the rolling fields. When he reached the top of the first rise he reined in his horse, gazing back toward the only home he had ever known. Martha and Caleb stood together watching him, as he knew they would, until he was out of sight.

The house that his father had built, the other homes and the church with its small white steeple — David drank in the sight of his village, savoring it as a thirsty man would his last drop of water. With a final wave to his mother and father, David turned his horse toward the east where the City, glowing like a smokeless fire in the night, awaited him beyond the River.

Chapter Sixteen

A Bridge From Hell

Starr could see and hear, but it was not the same.

Somehow she was aware of *two* distinct scenes — both of them clear, even though they were separated. It was not like watching two television screens at the same time. To do that one must shift the vision from one to the other. And when one watched the first set, the mind automatically filtered the signals from the other.

No, she was simultaneously aware of *two* scenes, and each drama came to her with a startling clarity — sharp and defined. *How can I think of two things at the same time — or see two things at once?* The thought arose in her mind, and she remembered a passage from an ancient novel, *Moby Dick*, which marveled over the fact that the sperm whale's eyes do not look straight ahead, but are located on the sides of his head, so that his brain receives two completely different sets of sensations.

One of the scenes was set in a hospital

room. There was no mistaking the white-coated men standing beside a bed, nor the patient with tubes running from the nose and other orifices of the body. She could distinctly hear the humming of machines hooked to the still figure by plastic tubes, and she could see the tiny dot that moved across the glowing green monitor, rising to a small mountain peak suddenly, only to drop sharply.

That's a heartbeat Starr knew, though her thoughts were not in words so much as in impressions. *This is an intensive-care ward.* She "looked" around the room, but not as an observer tied to one spot. She was able to see the faces of the two Healers in white coats, and at the same time the man who stood across the bed facing them.

Why — that's Richard Xi! The thought seized her, and then she looked at the pale face of the patient, and realized who it was covered with a white sheet.

That's me, she thought, and the knowledge came not as a terrible shock, but as a confirmation of a truth she had already possessed.

Then one of the Healers, a short, heavy man with a short gray beard said, "We don't fully understand it. As I told you yesterday, she seems to have no serious brain damage — in fact, very little injury."

"Then — why doesn't she wake up?" Xi's

voice sounded desperate and his face was gray with strain. "You must have left her on the Blood Siphon too long."

"It's part of the standard procedure for everyone sent to the Healers. It's monitored very carefully."

"Not carefully enough, apparently." Xi paced back and forth next to the bed.

"Well — I wish we could give you a better answer. We've done a great many tests — and to be truthful with you, my colleague and I are just not certain what to do next."

The other Medical, a very tall man with a full head of black hair and dark intense eyes, spoke quickly. "This sort of thing is not altogether unknown. It may be a combination of loss of blood and shock — and she's doing well. I think we ought to do nothing at all for at least twenty-four hours. In my judgment she'll come out of it very soon."

"She must come out of it!" Xi appeared barely able to constrain his rage. "If she dies, we'll have a martyr on our hands. That would only promote the spread of —" Xi caught himself in time. "I have to go before the White Council soon. I need good news to tell them."

Starr longed to return to the Fields — to see David and his family — to see again the rivers and fields and the forests. But there was nothing to be done in the usual way. She knew that, but she didn't know *how* she knew it.

She could see and hear, even smell the antiseptic of the intensive care room — yet she was not there at all.

Where am I then? she wondered. She was mildly surprised that fear was not piercing her — and almost clinically the thought was in her mind *I must be dead.* But even that thought didn't frighten her. To some extent, it was as though she were encased in some invisible bubble, with her senses intact so that she could see and hear — but the sights and sounds from the other world had no power over her emotions. *I ought to be more concerned for that girl on the bed with all the tubes.* That came to her — yet she felt no compulsion to create a pity or compassion for her body that lay so still before her eyes.

That was one world that she was aware of, and it was not exotic or strange, composed as it was of familiar figures and devices.

The other "world" or "scene" was no less ordinary, for it was her own bedroom. But with one difference: she was not looking into the room, as was true of the hospital room, but was actually *inside* the room — but in this vision, she was not looking on, not seeing herself, but was inside her own body.

She was lying on her bed with her eyes

closed, and as she lay there the same tinkling music seemed to be in the air, only faintly. The tufts of the chenille bedspread tickled her palms and the calves of her legs. She could smell the faint odor of the cosmetics which she kept on her dressing table.

Even though her eyes were closed, she was aware that someone was in the room with her. As with the other setting (which she was still aware of) she was not alarmed by what would ordinarily have been a frightening situation; rather she was curious.

But Starr was also seeped in a warm comfort, a state so peaceful that she had no desire to open her eyes for what seemed like a long time. She lay there aware of the room, aware of the girl on the hospital bed — but was reluctant to open her eyes and discover who was in the room with her. *It must be David,* she thought once without being greatly concerned.

Finally a voice came to her, a man's voice, saying, "Starr Omega, open your eyes."

She slowly opened her eyes and saw at once the figure that stood beside her bed. He wore a simple white coat with large pockets and loose-fitting white trousers. It was his eyes and voice that she never forgot — and yet even those two qualities she could not describe to her own satisfaction.

He was rather tall; his face was neither

young nor old. Like some orientals, he could have been any age from twenty to fifty — nevertheless, there was such strength in the lean face that age did not seem important. He had very dark eyes that appeared deep as wells, filled with a compassion and wisdom that seemed to overflow as he stood there watching her. He was dark rather than fair; his mouth full and mobile.

Starr sat up in bed, then stood to her feet. "Do I know you?" she asked.

"Not as well as you will later," he said, and his voice was quiet and low. "But I know you, Starr."

She listened, then stared at him. "I'm dead, aren't I?"

"No, no!" he protested at once, and a smile touched his lips as though he entertained an amusing thought. "No great wonder you should think that, I suppose." He studied her carefully, then said, "I've been sent to give you a chance to help, Starr."

"Help? Help who?"

"Yourself, your family, your friends — and many people you've never even met."

She was watching him, and going through her was that feeling that comes when you *know* you've met someone — but have absolutely no idea where or when. "I do know you," she insisted. "We've met before!"

He smiled and laughed audibly. "Well, I suppose that's true — in a way. But for now, it's best that you don't know much about me. Suppose you just call me Goel." A smile touched his lips and he added, "That's an old Hebrew word. It means 'One who stands between me and another.' Are you afraid of me?" he asked suddenly.

"N-no, I don't think so, Goel," Starr said. "But I'm confused. This is my room, isn't it?"

"Yes."

"But — I'm in the White Tower? I've been sent to the Healers by Richard Xi?"

"That's right." He came closer to her and asked, "Starr, do you remember the dreams you've had?"

"Well — I've been having bad ones and good ones."

"I know. I've been coming to you for a while, asking you to go through the mirror."

Suddenly she *was* afraid, for the first time. She cast an agonizing glance at her mirror, then took a step back, anxious to get as far away from the still figure as possible. "No! I won't do it."

He regarded her without moving, but said, "No one is going to force you to do anything, Starr — well, at least *I'm* not." Then he added, "There are others who might try to control you."

"Others?" she cried, and looked around the room. "What others?"

"That's what you will never know — unless you're willing to trust me." He held out one hand suddenly, and compassion was in his voice as he urged, "Starr — you've got to choose! I know others have been trying to control you. But if you'll let me be your guide, I think you'll be able to understand what is going on — not just in your own life, but in the world."

She stared at him, and whispered, "Goel, you want me to go through that mirror, don't you?"

He shrugged and his face was very sober. "Not *that* mirror, Starr." He waved his hand toward the oval mirror beside her bed, adding, "That's just an ordinary mirror, made of wood and glass. If you tried to go through it, you'd probably get some severe cuts — and there's nothing on the other side except the wall!"

She stared at Goel, fear rising in her heart, but the sight of his face and the assurance in his dark eyes caused her to ask, "Then why do I keep dreaming that you want me to go through the mirror?"

"It's just a way of putting it, Starr. Some things can never be put in scientific terms. Love, for example." He studied her and there was something about him once again that looked familiar, though she could not have said how that could be. "You can't say, I love you 354 times 126, can you? Of course not! That's why man has poets, to talk about love in ways that most people can understand. And the world you know, Starr, isn't the world at all — not the real one."

"I don't understand."

"Few people do," he said sadly. "But you know something like it. Because you're here in this room with me — and at the same time you're in the room where your body is, in the hospital. Isn't that so?"

"Y-yes, but that's because I'm — I'm in some sort of a twilight zone!"

"Exactly!" he exclaimed. "A twilight zone is the best way to describe what you're experiencing right now. But most people are aware of only one world. They think little about another world that's just as 'real' as the one they live in."

Starr stared at him, then asked, "But why do I have to go through this — this *mirror*? What is it?"

"It's a way that will enable you to see the 'other' world, Starr — that world that only a few poets and a larger number of saints are aware of. A mirror — a *real* mirror — is only a piece of glass with a silver coating on one side. When you look into it, you see a reflection of things. But if all the backing were removed you'd see what was *on the other side of the mirror.*" He nodded and added, "What you are called to do, Starr, is learn to see the world as it is — not just reflections."

"What will happen to me if I do as you say — go through the mirror? Will it hurt me?"

"It might." Goel's face was suddenly stern, and he said evenly, "There are no guarantees, Starr. Those who would do good in this arena must pledge their heads to heaven."

The phrase caught her attention. "Pledge my head to heaven? That means I might *really* die?"

"Will you come with me through the mirror?" he asked, his steady gaze locked on her face. "We never beg, you know, for only volunteers would be any good at this sort of thing."

She wanted to ask who "we" might be, but knew suddenly that he would give her no answers. She understood that somehow the interview was at an end, that either she must go with him through the "mirror" — whatever that was — or go back to the room where her body was in a hospital bed.

She had the feeling that all of her life she had been making her way down a very broad path, in the company of many people, and now she was being asked to step off that comfortable pathway onto a rough, dangerous way that led to an obscure destination. Never had her life seemed so attractive and secure, and she longed to turn her face and flee from the one who stood waiting with longing in his dark eyes.

Though she was not in the body, she felt the weakness that clings to the flesh — the trembling in the knees, the oppressive weight on the chest, the dryness of mouth. She looked into Goel's eyes and pleaded, "If it were only a *physical* danger — it wouldn't be so bad. . . . !"

"No, that's the *least* part of courage," he agreed. "That's because with physical danger, you can only lose your life. But we're talking about something much more valuable than that, Starr. You're risking your soul. I tell you plainly that many who have gone through a 'mirror' such as the one I set before you, have lost everything." Then his voice grew stern, "Choose at once, Starr! There is little time."

She stood poised, ready to flee yet unable to deny the impulse that rose in her. She seemed suddenly to hear her mother's voice saying, "Don't be afraid, Starr! Trust in God!" And she knew, somehow, that wherever her mother was, she was aware of her plight, and was begging her to keep on.

"All right," she said unevenly. "I'm afraid

— but if it will help, I'll go with you."

He smiled, and his eyes were filled with a golden joy that shone forth like sunshine. "You will be afraid many times, Starr," he said gently. "But that is not the test. It is not those who have no fear who are able to serve God — but those who go on into the darkness holding their fear in their hands — as you do now." Then he said, "Take my hand, Starr — and I will take you to where you must go."

She took his hand, and it was warm and strong. "Are you real?" she asked curiously, ignoring her fear. "I mean, I suppose you're an angel, aren't you, Goel?"

"Certainly!" he said firmly. "But don't be expecting all angels to look like *me.* We pretty much don't *look* like anything, Starr. We take on whatever form is least likely to startle people." Shaking his head sadly, he added, "But you'll remember in the Bible whenever we appear to one of you of flesh and blood, the first thing we have to say is 'Fear not!' "

"You do look familiar," she insisted.

Then he tightened his grip and said, "We're going through the mirror now — hold my hand. And I'll be holding to you, Starr. Not to your hand, but to the real part of you. Just remember — you've been looking into a flat mirror all your life. All you've ever seen are reflections. Now you're going past that, into the realm where *reality* is present."

And then it happened.

The room faded into a mist and there was a sudden sense of space — immense and vast. Her hand was held tightly, and she knew that it was Goel's way of giving her courage, but she was no less aware of tremendous activity as the last vestiges of her room faded and she found herself in a setting that she could not have imagined.

She was in space, but it was not the cold, empty space reported by science and astronauts. A million glittering points of light greeted her — not icy dots, but all alive with a force that pulsated through the whole cosmos! They were not fixed, she became aware, but were quivering, moving across the sky in a magnificent dance, stately as a minuet, yet joyful as a polka! She had never seen such life; it made what she had seen of earth look dead and dull, and she whispered, "Oh, Goel, how beautiful!"

"Yes, it is. He hath made everything beautiful in its time," Goel said. "You are seeing reality, Starr. All this is the way things *are* — the signature of the Creator who delights in beauty, who makes all things from that galaxy over there trembling like a million diamonds to the segmented earthworm that burrows blindly through the clods of earth. He made them all!"

For what seemed like a long time Starr drank in the myriad burning stars, the moons, the suns and their planets, and knew that no astronomer had ever seen such things — but that she had passed through the mirror that hid the depths of the great creation above the earth.

Then she exclaimed, "What's that?"

"That's your home, Starr," Goel said quietly. "That's earth — the crowning glory of God's creation. It's the one He made for your race — and the one He sent His Son to redeem."

Starr gazed at it, and the beauty of it made her eyes burn with sudden tears. She whispered, "I remember how a man from OldWorld named Milton described it."

"One of God's saints, John Milton," Goel said. "What did he say?"

She quoted the lines she had always loved:

Far off th'Empyreal Heav'n, extended wide
In circuit, undetermin'd square or round,
With Opal Tow'rs and Battlements adorn'd
Of living Sapphire, once his native seat:

And fast by hanging in a golden Chain
This pendant world, in bigness as a
  Star
Of smallest Magnitude close by the
  moon.

Goel looked into her eyes and smiled. "It is like that, isn't it — a pendant world hanging in a golden chain?" Then he said gently, "But now I must show you something else. Do you remember in Milton's poem the lines that come just before those you've just quoted?"

"No, I don't remember."

"Come then — and I will show you another reality — the dark side of the mirror."

And they were moving, not in space so much as in another dimension that was not space or time. She became aware again of tremendous activity, everywhere and constantly. All she could say later was that it was like a wind — not a breeze, but a wind that whipped from every direction at the same time. But these were not winds created by the forces of nature, but rather *energies* that crackled with what seemed to be violent electrical charges. Or, she thought, it was like an ocean, where the waves contended with one another, bringing chaos to the sea.

"It's — like a *battle!*" she cried aloud.

"Yes, it is a battle," Goel answered. "A

battle that would put Waterloo or Hastings to shame, Starr." Then he seemed to pull her away and they moved to a point where convolutions seemed less — in fact, for the first time, space seemed to be enormous and quiet. She saw again the earth, like a huge pendant, and delighted in the green continents and the dark blue of the seas. Again she said, "It's so beautiful!"

"Yes — but there is a danger, Starr," Goel said quietly. "Not all want it to remain beautiful. There are those who want to use it, to make it a huge ugly dungeon for your people. Look — you see that dark line?"

Starr turned and saw — or seemed to see — a filmy track or line that was attached to the world. It was darker than the blackness of space, standing out against the ebony as an even blacker thread. And as they moved toward it, the darker it got. And then she saw it more clearly.

"Why, it's some sort of track, isn't it?"

"Something like that. It touches earth, but look where it leads to."

Starr saw that the ribbon of darkness came from space, that it seemed to emanate from some part of space so dark that the rest of the sky was a foil, setting off the utter darkness of that spot.

She looked closer and said, "Something is

going from that dark place out there to earth!"

"Yes," Goel answered. "Now, I will quote you a few lines of your poet Milton. In some way that I do not understand, he *saw* this that you see now and wrote about it. You will remember that Satan had escaped from Hell and fought his way through the Abyss until he finally arrived in space. Sin and Death had followed him out of the Pit, and in a few lines Milton described how Satan came to your home. He said:

. . . Sin and Death amain
Followed his track, such was the will of
  Heav'n,
Pav'd after him a broad and beat'n way
Over the dark Abyss, whose boiling Gulf
Tamely endured a Bridge of wonderous
  length
From Hell continued reaching the
  utmost Orb
Of this frail World —"

"A bridge from Hell!" Starr cried. "I remember that! A bridge that ran right from Hell to earth!"

"Yes," Goel's voice was filled with pain and anger. "And why is it there? The next few lines tell us why for they say that this bridge is the means: '. . . by which the Spirits perverse /

With easy intercourse pass to and fro / To tempt or punish mortals, except whom / God and good Angels guard by special grace.' "

"Now you know my task, Starr," Goel said quietly. "I am one who is set to guard you with special grace."

"A guardian angel?" Starr breathed in wonder, staring at him. "I never believed in those."

"But you didn't believe in God — not really." She dropped her head and tears stung her eyes. "Don't cry," he said gently. "You believe now."

"But not everyone has a guardian angel appear to them. It doesn't seem *fair!*"

"All is fair with God," he rebuked her gently. "Shall not the Judge of all the earth do right?"

"Sorry!" she said at once.

"Yes, I know you are — but you see, Starr, not all have the good fortune to see angels — but neither are all called to such a hazardous task as you are."

His words frightened her, and she asked in a small voice, "I'm not very strong, you know. Are you sure there's not some mistake? Maybe it's somebody else."

"We keep very good records," Goel said dryly. "You are the one who has been chosen for this task. Others will have their own calling. Now, let me show you one more thing. . . .”

He waved his hand and she saw that streaks of light were running across the surface of the world. Some of them were mere dots of light; others were much larger. They seemed to come and go constantly, and she asked, “What are those, Goel?”

“It means a battle is going on in that spot. Milton wrote about the war in heaven, Starr, but that war is over. The battleground is not in heaven or in some distant galaxy — but on planet Earth. And the battle is joined. Souls are dying, and the enemy is strong.”

Starr studied the globe of her home planet, then asked, “Why are there more flashes of light in some places than in others?”

“Those are areas where the spiritual war is raging most intensely. In those places witchcraft is everywhere. So the Servants of the Lord must be there to protect the people of God.”

“All those flashes are angels fighting against demons?”

“No, not all are angels,” Goel said, then took her shoulders and looked down into her eyes. “Those flashes of light you see are for the most part believers who have learned how to do battle with the powers and dominions in the service of the Evil One. Come, and I will show you something."

They moved through the reaches of space, and earth seemed to swell and grow. The face of the earth had changed since the days of OldWorld, but Starr had once seen a Classified Document in one of her searches. It showed the world as it now was. They were over the North American continent, and she could see the River cutting its way through the United States. Closer still, and she cried out, "Look! There's where I live, Goel! That's where the City is!"

"Yes. It's your home, Starr — but not all those lights are from the Dome."

She peered closer and the land seemed to rush up, and she saw one area where the flickers of light were clashing with dark splotches. She trembled and whispered, "What is it, Goel?"

"It's the battle that's taking place over you, Starr."

She stared at the flashes, which were like far-off lightning, and whispered, "Why is there fighting over *me?*"

"The enemy knows you have been with us — with the Servants. They know that means you have been chosen to do a work for God. And that is, to the Enemy, a call to battle. The Evil One himself disputed with one of us about the body of Moses. So now there is a mighty gathering of Dark Spirits who are determined that you will *not* return to your body."

"But — what will they do?"

"They will kill you if they can, Starr." The face of the Servant was grim, and he asked, "I will not deceive you. The risk is great — and I cannot help you."

She was stunned by this. Somehow she had assumed that the Servant would make all decisions, and if necessary do all that had to be done. Now she saw that it was not going to be like that — and she cried, "But — why not?"

"I do not know. It is the will of our Lord that the battles must sometimes be fought by human beings." He seemed to meditate on this, was silent for a long time, and they stood there watching the fierce conflict that was taking place beneath. Then he said, "There is a mighty warrior of God down there, Starr. All alone against a phalanx of the Evil One's most powerful spirits."

"Just *one* — against all those!" she whispered. "It seems so hopeless!"

"Do not say that!" he urged. "It is not hopeless. But now — you must decide. Will you be the servant of God, Starr?"

"But, I'm no angel!"

"No, God has many servants — but that is

not your problem. The time is this, Starr — will you go back to your home? And will you be a servant of the Lord Jesus — no matter what the cost?"

She was stunned, and could not think. Her mind reeled and she longed to refuse. Yet — something was *pulling* at her — something from that maelstrom of battle far below. And then she realized it was a voice, a familiar voice!

"Come back to us, Starr! Come and serve the Living God!"

"Why — that's the girl I saw in the Rim that day!" she exclaimed. "Is *she* the warrior, Goel?"

"Yes — and a more valiant one the Lord scarcely has in all His host!"

Suddenly Starr felt a surge of hope — and of joy. "Yes! I'll go back, Goel!"

"It may not be possible. The girl is under a terrible strain. If she falters, you will be at the mercy of those vile creatures you saw on the bridge. What will you do then?"

Starr was trembling in every nerve, but the call was strong and clear now — *Starr — you must come back!* She lifted her head, her eyes filled with tears, and she sounded much like her grandmother at that moment.

"Yes! Yes, I'll go back!"

The Servant said with a great victory in his

eyes, "To God be the glory!"

They stood there quietly, then she asked, "What will happen now, Goel?"

"You must go back. It will not be easy, but you must trust the girl. She is standing at the bridge where the Evil Ones cross, turning them back. She will not falter."

"And when I get back, will you be there to tell me what to do?"

"Not for a time," he said gently. "It is best that you learn how to be a servant, not from one of the High Ones — but from one of your own nature. We angels know much," he said thoughtfully, "— but none of us are redeemed. That is the *glory* of your race, Starr."

"But I won't know what to do!"

"You will not be left alone. A Teacher will come. He will help you."

"Who is he?"

"That you must find out for yourself," Goel said. And then his voice grew stern, "Be careful, Starr! Many false teachers will surround you. Some will seem so good and wise, but will be wolves in sheep's clothing. Your Teacher will make himself known — but you must choose him."

"But *how?* I don't understand," Starr protested.

"I can give you three signs," Goel said.

"Listen carefully and do not forget them. First, he will come by a name you do not know. Second, he — or she, as the case may be — will not be accepted by your family or your friends. And third, he will tell you something about yourself that you do not know — that no one on earth knows."

She moved her lips silently, memorizing them, and asked, "But that's so *vague!* Can't you —"

"It's time for you to go home, Starr," the Servant said firmly. "Are you ready?"

Starr swallowed, then looked down at the dark forms and the flickering sparks of light. Desperately she wanted to go with Goel, to do *anything* but go down and thrust herself into the fierce battle below — but she knew the time had come.

"Yes, I'm ready," she whispered.

His hand closed on hers once more, and he said as she began to rush through space, "Do not fear, my child. You are a Chosen One — and the Lord Himself will be your shield!"

Her ears roared as a wind rose up from earth, swallowing her cries to the Servant who was no longer there, the clasp of his hand gone.

And as she rushed down toward the dark scarred earth below, she heard two voices — one was the girl calling out in pain. The other was a hoarse, formless voice that suddenly tore at her like a physical blow. And then she felt the pain as dark, spectral things ripped at her, and then she cried out to God as the darkness closed in on her.

Chapter Seventeen

Warrior

David tethered his horse and waited at the edge of the darkness. He could see the small domes glowing like huge yellow eyes at the tops of the twin black towers, marking the gate through the laser wall. The wind from off the River, just over the next rise, was damp and cool, bringing to him the sound of croaking frogs. He thought of his home with its own River at the foot of the wooded bluff.

"Come quickly into the light," rang out a voice from a speaker somewhere near the entrance. "When this message is complete, you will have ten seconds to pass between the columns. Do not hesitate. You have only this one opportunity." A harsh buzzing sounded the end of the message.

David had expected Will to appear — instead he heard only this metallic sounding voice in the darkness. He glanced at his horse, knowing that once beyond the gate, there was no turning back — there was no recrossing of the laser wall. Sprinting across the field of light, David passed between the black columns and paused at the edge of the darkness on the opposite side of the fence.

"This way!" The voice, coming from the direction of the River, belonged to Will Sigma. David walked through tall grass up a gently sloping hill toward the sound. Halfway up, he saw the silouetted form of a man standing on the summit. "I'm glad you had the good sense to wait that first night, before you started your journey," Will murmured in the freshening breeze. "One of the Special Units was on alert in case anyone followed us. Everything's quiet now."

David bounded easily up the hill. "Is she all right? Where have they taken her?"

Will sat on a crest of rock, motioning for David to sit next to him. "The worst possible place — the place I thought they would — the White Tower!"

"What's that?"

Will explained briefly the structure of the City and what awaited David there. "You'll need some help and there's precious little available under the Dome."

"I've got to get her out of there and across this River," David said gravely, an intense light in his dark blue eyes. "They'll never take us once we make it to the Fields."

Will took a small light from his coat pocket,

switched it on and sat it on the rock beside him. Unfolding a sheet of paper, he placed it on the rock next to the light. "This would only be a waste of time for anyone else, but you just may have a chance to pull this off, David. I can get you inside the City — put you in touch with one man you can trust. Now here's what you have to do —"

In a few minutes Will pointed back beyond the gate at the bottom of the hill. "I'll tether your horse and Starr's down in that stand of willows next to the stream tomorrow night and for two nights afterwards. If you haven't made it by then —"

"You have Starr's horse?"

"The Guards keep all the horses." Will switched off the light, slipping it inside his pocket. "No one has ever escaped from the White Tower, David — ever!"

David glanced at Will, then looked down the hill to the lighted barge landing. "I don't go in my own strength, Will, but in the name of the Lord of hosts that delivered Israel."

The long grass on the hillside waved gently in the wind like the surface of the sea. Far beyond the River, the great Dome glowed coldly against the night sky.

There was almost no traffic as David traveled along the dirt track that ran from the City to the River. It had been worn down over the years by boots and horses' hooves until its hard-packed surface was several inches below the ground on both sides. David ran easily along its edge, melting into the underbrush with his gray cloak about him when he heard someone coming. *They've never had to practice stealth,* he thought gratefully.

A quarter of a mile from the City, David climbed to the top of a tall beech tree to get his bearings. The light from the Dome showed him the huge main gate where several Guards, their horses grazing contentedly, slouched around a rough building of weathered lumber. He could barely make out the smaller track skirting to their left. Moving at the edge of the woods, he continued past the small gate where Starr had entered the Dome in the carriage. Two hundred yards further on, after no more than a cursory glance around, he located it.

Brushing back some leaves, David saw the dull metal ring. He lifted on it, the small trap door swinging easily upward. A roughly welded ladder descended into a narrow dark void. With no hesitation, he swung easily into the opening, closing the door over him. The metal rungs of the ladder were cold on his hands as he descended into a darkness that made the night he had just left look like noonday.

* * *

Lido slid the panel back one inch, peering out over a bale of hay into the gloom of the stables. The horse directly below him, a spirited pinto, smelled him at once and began neighing and shaking his head. After he had quieted, Lido slipped through the panel, closed it behind him and scurried over to the edge of the loft behind several stacked bales of hay. In the back corner of the main stables below him he could see the door to Michael Kappa's small quarters.

Scampering down the ladder from the loft Lido ran across the hay-strewn floor and opened Michael's door. As he stepped into the cramped room he froze in terror. Talking to his friend was a tall man in a gray cape. He had long black hair, dark blue eyes and a look about him that in no way resembled the people that lived under the Dome. It spoke of open spaces and sunshine and rain sweeping across the endless green Fields. Lido felt this, although he didn't understand why as he had never seen any of those things.

"Come in, little friend," Michael assured him. "This is David. He's come from the Fields."

Lido looked up into the face of the tall stranger. His elflike appearance and his life in the underground and hidden places of the City were as different from David's as two people could be, but he saw something that transcended all the outward differences between them, something that made him feel he had known David for so long a time that he could trust him. "Have you really come from the Fields? Are there trees and rivers?"

David liked this small, inquisitive intruder and felt immediately the kinship the two of them shared. "Yes there are — and much, much more," he laughed.

"Can I go there with you?"

David sat on one of the two straight-backed metal chairs, resting his elbows on his knees. "I think perhaps you can, my brother. You'd dearly love it out there. But first we have work to do and Michael says you're the man for the job."

Lido looked puzzled. "Job?"

Michael sat down in the other chair. "You can have your usual place." He motioned toward his bunk.

Lido bounced onto the bunk, sitting in the middle of it with his legs crossed.

"David wants you to help him get someone out of the White Tower. Someone they took from the Fields." Michael waited for Lido to bombard him with questions.

"Starr Omega."

"You know her?" David asked, astonished at the response.

"No," Lido answered, bouncing lightly on the bunk. "But I've seen her."

David leaned forward eagerly in his chair. "How did you know it was her?"

"I listened to them."

"You were in the White Tower?" Michael's eyes narrowed in doubt as he stared at the little man who had such a childlike innocence about him. "No one goes there."

"I do."

David gazed intently into Lido's bright eyes. "Will you take me there? Help me get her out?"

"That's why I came here," Lido answered excitedly. "Philea and I were going to get her out ourselves. We needed Michael to help us get out of the City."

"Well then, David makes four of us in on your little caper," Michael mused. "The three of you bring her back here to the Camp and I'll have things ready to get us out of the Dome."

"You're going with us to the Fields?" David was encouraged by the thought of someone like Michael throwing in with them.

"Since I became a Christian, it's all I've thought about," Michael smiled. "This is an answer to prayer."

"I'll have to get another boat," Lido piped in. "I just have the one for Philea and me."

David was perplexed. "Boat? Why in the world would you need a boat inside the City?"

It was like something out of the imagination of a demented mystic, this underground world beneath the City. The deep rumbling drone of the massive generators overlaid everything, making conversation impossible unless you were close to the person you were talking with. The foundations of the buildings looked like giant steel pylons turned upside down, extending from the underside of the buildings themselves. They were attached to huge metal bases resting on a series of giant springs as protection against earthquakes.

Steam and water pipes of every imaginable size ran vertically, horizontally and diagonally in an incomprehensible maze. The sibilant clouds of escaping steam added counterpoint to the heavy drone of the generators. Shafts of pale yellow light dropped from far above through the grates of the streets, creating the impression of perpetual twilight. Through this spectral world, beneath the stark spires and canyons of the City, flowed a River.

David stood with Lido next to one of the gigantic concrete pillars that supported the City — that provided the stability for all the other networks of systems that sustained it with air and light and warmth. The River flowed before them, warm and languid, and incredibly clear, between the vertical walls of a concrete canyon.

David glanced down at Lido. "Why is the water so clear? There's not even any debris."

"The Rippers." Lido answered as if that should be enough information for anyone.

"The what?"

"The Ripper fish," Lido explained. "They eat anything that gets into the River. I don't think there's anything they don't eat — even if they can't digest it. Nothing seems to hurt them."

David was fascinated. "What do they look like?"

Lido picked up a small chunk of concrete that had broken away from one of the columns. "Watch."

David watched Lido toss the chunk over the concrete bank of the River. As it hit the surface of the water forty feet below, a gray fish, three feet long and shaped like a fleshless barracuda, streaked toward it from a small side channel. Catching the concrete in its mouth, full of razorlike teeth, it chewed and shook its head viciously. Then, like an aquatic shadow, it disappeared back into the dark tributary.

"They keep the River clean," Lido smiled.

"I'll remember that," David remarked, looking up and down the River at the plethora of small channels that either emptied into or flowed away from the main channel.

Lido sat on the concrete bank of the River watching it flow southward away from the center of the City. A hundred yards downstream, the River dropped into a deeper underground channel in the earth, disappearing under the edge of the Dome. The sound of the River's descent into the subterranean darkness was that of a small waterfall.

"When will Philea come?" David asked patiently, learning to adjust his pace to Lido's.

"Perhaps an hour — maybe less. She's been in fasting and prayer for two days. Now she's gone to the Fringe to locate one of the special gowns that the Physic Advisors wear," Lido answered absently. "You can find just about anything in the Fringe if you know where to look."

"I don't understand why she's going with us." David walked over and sat next to Lido, looking down at the river, shivering slightly as he thought of the Ripper.

Lido smiled benevolently at him. "The Physic Advisors are the only ones who can remain in the room at all times with patients under the care of the Healers."

A shadow flickered in David's eyes. "Is it so necessary to have someone with Starr continually?"

Lido answered in a grave tone. "The White Council hates Christians because they fear what they don't understand and because they don't have the absolute control over them that they do the other citizens under the Dome. They think this freedom will undermine their power. That's also why they'll keep Starr alive, as long as they think they can get her to publicly disavow her beliefs."

"Then why do we need someone with her — if they don't plan to kill her?"

Lido stared out across the River at the endless maze of concrete columns, steel pylons, pipes and other structures of abstract and infinite strangeness. His face seemed to change as he spoke. "There is a group in the City that sees Starr as an even greater threat to them than the White Council does."

David stared as Lido's face seemed to become that of an old man. "Who are they?"

"The Coven. And they see the true danger as allowing Starr to remain alive at all — even for a short time. They know her threat to them is not of this world." Lido's face brightened. "That's why we need Philea. She knows how to fight them."

"She must really be something," David declared.

"Oh she is," Lido agreed.

Across the River, in the upper levels of the towering shafts of concrete and steel, a small figure moved catlike across a beam spanning an open space. He hooked a belt around one of the supports and slid down it to a platform where he began some sort of repair work, using tools from a bag attached to his waist.

"Who's that?" David asked, fascinated by the agility of the small workman.

Lido glanced upward. "Oh, he's just one of the Shadowmen. They keep things in working order down here."

"They come all the way down here every day?"

"No." Lido watched the man as he made his repairs. "They live down here."

David looked about him at the absolute lifelessness of the surroundings. "Here — all the time?"

Lido smiled benignly at David. "It's better than the Rim. At least down here no one bothers them."

"From what I've seen of the City, they must be under the scrutiny of someone."

"Oh, they were once — a long, long time ago. Then, over the years, their overseers came down less and less until they more or less forgot about them." Lido waved to the little man on his high, precarious perch. He put his tools away and waved back. "The Shadowmen found out if they kept things running well, no one bothered them. I doubt there's anyone left who's even responsible to check on them anymore — and that's the way they like it."

"It must be a lonely life," David murmured, watching the repairman scramble back up the support.

"Oh no!" Lido answered quickly. "They have their families with them."

Shaking his head gently, David stared at the Shadowman as he disappeared into the murky recesses of the City's substructure. "The resilience of the human spirit never ceases to fascinate and amaze me, Lido."

"They're just doing what it takes to survive like the rest of us," Lido remarked staring back into the River. "And they're very good at it. I think they have access to every building in the City from down here. That's how they get their food and anything else they need. They take just enough to survive so no one will come looking for them."

"Did they show you how to get into the White Tower?" David asked abruptly.

Lido smiled, nodding his head. "And a lot

of other places too. It's fun to go with them sometime."

Looking about him at this forbidding and complex world with its own River and aerial pathways, David said, "I guess it's an interesting life down here — and a safe one."

"Maybe not for much longer," Lido remarked cryptically.

"What do you mean?"

"Christians trying to get away from the Peacemakers are coming down here — more and more of them. I believe they'll soon find out about this place."

"What makes you think that?" David asked, turning his attention to Lido.

"The things I've heard in my journeys under the City — especially in the White Tower and the Department of Adjustment." Lido frowned. "It's going to be very bad."

David heard the swish of soft shoes on concrete and turned to face her — and his spirits fell! Thin and pale and fair-haired, she was more girl than woman, having a fragile look about her as though she should be protected behind glass. She wore a diaphanous gown sprinkled with gold that gave her an ethereal air. Incongrously, she carried a heavy canvas bag over her shoulder. In her free hand, she held a small blue cloth bag with a drawstring.

Seeing her in the stark and murky environs

beneath the City, David felt she was a vision rather than a person, expecting her to disappear any second.

"I'm glad to meet you, David. My name is Philea."

With the first soft sound of her voice, the vision became flesh and blood.

"How — how do you know me?"

"I could never mistake you for anyone who lives in the City — and neither will anyone else." Philea handed him the canvas bag. "That's why I brought you a present."

As David took the bag, he gazed steadfastly into Philea's light blue eyes and began to sense that there was far more to her than her appearance suggested.

"It's perfect!" Lido exclaimed with sudden insight, before David had opened the bag.

"How do you know?" Philea smiled.

Lido shrugged knowingly. "There's only one way David could go undetected in the City."

David pulled a shiny helmet and the white uniform of a Peacemaker from the bag. "You expect me to wear this?"

"I think you'll cut a dashing figure in it," Lido teased, trying on the helmet.

David laughed at the sight of the little man in the huge white helmet with its black visor. "Maybe so, but I think I'll wait until we get to the White Tower to put it on."

"We'd better be going," Philea urged. "It's a long trip and we have very little time to spare."

The three of them made a strange amalgam: David, wearing his buckskin shirt and boots with his weapons slung on his back; Lido, tiny and agile in his gray coveralls; and Philea, in her shining gown; climbed down the steel ladder to the small ledge at the edge of the River. Tied to a pipe that ran along the concrete wall forming the River bank, were two canoes made of a black plastic material. They were eight feet long and narrow, having no seats in them.

David looked at the unsteady craft, glancing about for a sign of the Rippers. "Are you sure they're safe?"

"No," Lido muttered, a wry smile on his young-old face. "But they're all we've got."

David dropped the canvas bag into the front of the lead craft, climbing carefully into it. He positioned himself on his knees and took up a small gray paddle made of the same plastic material as the boat. Lido and Philea boarded the other craft with Lido in the rear handling the paddle. With the soft swishing of the hulls against the water, they began their journey through the perpetual gloom of this surrealistic world beneath the City.

Paddling with a smooth, steady stroke, David noticed a movement in the cellophane-clear water beneath him. Three Rippers had appeared almost magically several feet below the surface — one just in front of his boat and two others flanking it closely on either side. They were all six feet long or more and their bony frames shimmered in the dim light as though they were made of metal — the yellow eyes gleaming with a patient and malevolent light.

"You assured me that Starr Omega could be trusted implicitly, Bernard." Richard Xi sat behind his white desk in his formally cold office in the White Tower. "It seems your assurances are worth little more than Sheila Phi's vow of silence."

Bernard Alpha always preferred offensive tactics rather than defensive. "I merely provided you with irrefutable facts regarding Omega's performance, Richard. I admit to championing her cause, however, I had no authority to sanction it. That, my old friend, required your approval."

Richard Xi's eyes, dark as a ferret's, fastened on Bernard Alpha. "I believe we've reached an impasse, Bernard."

"It would seem so," Alpha replied.

"It's also appallingly obvious that we both

have a vested interest in the psychic remediation of Starr Omega," Xi continued. "The White Council frowns on anyone even remotely connected to a conversion such as Omega's."

Alpha clasped his hands together, resting his chin between the knuckles of his forefingers. "I've given this a lot of thought. I don't plan to be sacrificed on the altar of Omega's insanity. The Healers are bungling this with remarkable ineptness — even for them. With their reliance on the Siphon and the Pyramid of Darkness, Omega's imminent death is virtually certain."

"You have an alternate plan?"

"That depends on the information you give me. I've heard only rumors so far."

Xi leaned back in his chair, his close-cropped white hair glinting in the harsh light. "What have you heard? Maybe I can assuage the anxiety that rumors usually spawn."

Alpha smiled knowingly. "Well, rumor has it that there's to be a purging, or maybe we should call it by a more accurate name, an inquisition of the Crossbearers. My sources tell me this is to be under the direct control of the White Council."

"We prefer to call them Disciples of Treason, the name used when the first outbreak of this blight occurred. The Action of

Enlightenment is to begin in the not too distant future."

"Action of Enlightenment — I like the sound of it," Alpha grinned sardonically.

"Now that you're *enlightened* as to the current and rather precarious status of the Disciples of Treason, what are your ideas for dispensing with the bothersome Ms. Omega?"

Alpha propped his elbows on Xi's desk, but jerked back in the face of his menacing glare. "What I propose is to let the Coven have their unholy way with her. They have absolutely no tolerance for Christians. And, since there's to be this Action of Enlightenment, the need to have her publicly recant her beliefs would seem to pale in significance. Even I know members of the Coven have infiltrated the power structure, at least as far as the Healers."

"Perhaps you're right," Xi mused, studying Alpha intently. "Of course we couldn't overtly endorse such an action. That would be tantamount to barbarism."

"Of course," Alpha smiled coldly. He sensed then a sudden, almost audible thud, like a door closing forever in his heart. The gleam in his pale eyes became incredibly like that of the Rippers, moving silently in the shadowy depths a thousand feet below him.

Chapter Eighteen

The Coven

Her body tingled from the salt water bath, and she was shaken by an inner agitation as she joined the others who were assembled around a nine-foot circle of rope. The only light came from two tall candles which rested on a crudely constructed altar — and she was glad that the room contained no bright lights, for despite her boldness, it had been harder to remove her clothes and join the twelve other naked worshippers than she had imagined.

By the flickering candlelight, she could see that the others were paying little heed to her — except for Natas Molech. The darkness behind the eye slits of the iron mask seemed to come alive as he looked directly at her — and she had to exert every ounce of will she possessed to hold his gaze. His breath was a cold vapor through the mouthpiece as he then nodded to a middle-aged woman standing to his right. She at once turned and went to a low table where a very large silver cup rested, picked it up and took it to Molech. He lifted it high with both hands, and raised his voice in a high-pitched incantation in which the others joined from time to time, then raised the cup to the slit in the mask and poured the liquid in. He passed the cup to the young man on his right, who drank eagerly before passing it on, and one by one, all thirteen of the members of the coven shared in the cup.

When the newest member lifted the cup and took a swallow, she knew exactly what it was — a highly potent psychedelic drug. She knew this because she herself had smuggled it out of the dispensary of the ward where she was the Head Nurse. Molech had told her that he preferred belladonna or a brew made from the amanita mushroom. "But the mushroom brew takes time — and we have a crisis that will not wait," he had said.

After the potion had been swallowed by all thirteen members, there was a reading from The Book of the Shadows by a tall elderly man — that she recognized instantly as the executive assistant to a member of the White Council. By the time he was finished, the drug had taken hold, and as the ritual progressed she could not be sure if things were actually happening — or if they were creations of her brain inspired by the drug.

She swayed back and forth noting the items that rested on the altar with the silver chalice — a hazelwood wand and a small caldron. A brass censer of burning incense, a five-pointed metal star engraved with signs, a bowl of salt, a black-handled sword (which she knew was called "athame") and a length of cord that symbolized the spirit that unites all the elements.

Molech called on a young woman to come forward, and as she stood in the center of the circle, he gave her a ceremonial scourging, after which he allowed an opportunity for anyone to share a personal request with the rest of the group. Several spoke their requests, including an overweight matron of forty who pronounced a curse on her Loving Friend and seemed to take great pleasure as the rest of the members chanted a vigorous and optimistic anticipation of the result.

Molech said in his shattering voice, "For our newest member, I must explain that the requests we speak result in a combined thought wave, and this we call 'The Cone Of Power.' It is a neutral force which can be directed only by a head witch or warlock, and can be a healing process — but can also be directed against a person or group of people." His voice fell to a low pitch, almost a guttural snarl and his features twisted into a mask of hate, as he added, "Tonight we are come together to do battle against a threat to our Master!"

A low murmur ran around the circle, and the eyes of those in the circle seemed to glitter in the candlelight, the newest recruit noticed. She felt very important, suddenly, her spirit seeming to swell within her, and she could not tell if it were the result of the drug — or the fact that Molech had fixed his penetrating eyes on her expectantly.

Then he said, "We have been given the means to crush this new danger — and the instrument will be our newest member." His eyes glittered, and he said, "Orion — we will have the sacrifice!" The tall man whirled and left the circle. He was back in an instant with a struggling young goat held firmly in his hands. He placed it on a rough table that one of the others had produced, and Molech was there, a gleaming knife attached to his stump of a hand. He began to chant and the others offered a refrain from time to time. Then he suddenly reached out and slit the throat of the struggling animal.

Though she was a nurse and inured to the sight of blood, the sudden gush of crimson brought a quick revulsion to her throat. But the others all released a cry of victory, and then Molech was standing before her, his hand reeking with the hot blood. He suddenly reached out and she felt the leathery stump on her face. It seemed to burn her flesh — and she was filled with an emotion she had never known — a mixture of sensuous joy, a sense of infallible power, and the raking of raw terror across her nerves. She would have fallen, but two of the members caught her, held her fast, and she heard Molech say:

"You are the instrument of His Majesty's wrath! Go to the girl. She is being kept alive only

by the persistence of one woman — who cannot last much longer. She hasn't slept or eaten in almost three days! Your assignment — bring back to me a lock of the girl's hair and some fingernail clippings. And be quick! We must destroy this girl — for we know that she has been chosen to be used by the Enemy. Even my Masters do not know exactly how — but we are commanded to destroy her at once. Now — go!" Then he added in a voice that was hard as steel and colder than anything she had ever heard: "If you falter — you will find out that our Masters can be cruel to those who fail — or even draw back!"

She stumbled from the room, washed the blood from her face, dressed and managed to get to the carrier. Her hands were trembling so violently it took several tries before she could open the door, and when the carrier began humming along the track, she wondered if she could even remain erect. Rolling down the window, she took in huge gulps of air, and finally the terror subsided. As she sped through the City, she planned in her mind exactly what she must do when she was in the same room with Starr Omega, and her lips were curved by an exulted smile as she thought, "It will be easy. Who would suspect the Head Nurse of belonging to the Coven?"

"Really, Philea — you *must* go home and get some rest!"

Philea raised her head slowly, looked at the tall form of the Healer who stood beside the nurse. Philea knew how she must look, for she had seen her reflection in the bathroom mirror an hour earlier when she had bathed her face with cold water. Face pale as paper, eyes ringed with red from weeping and from fatigue, mouth pale and drawn tightly together, great dark blotches under her eyes which seemed to have withdrawn into the cavities, giving her a skull-like appearance.

Every nerve in her body cried out for rest, for she had not closed her eyes in sleep since she had come to stand guard over Starr. Her limbs, she knew, were so weak that she didn't dare try to stand up and face the three who had come to persuade her to go home. Instead, she took a tighter grip on the blue bag in her hands, and spoke in a voice that was firm as stone — though raspy with fatigue.

"I will stay until Starr comes back."

The Healer shifted uncomfortably at the words, for they didn't seem *right* somehow. *Until Starr wakes up* would have been more correct. He caught a warning shake of the head from the other Healer, who stepped forward and put his carefully manicured hand on Philea's frail shoulder, saying in a soothing voice, "Now, Philea, we don't want to have you here as a patient, do we now? We don't have enough good Psychic Advisors as it is. Go home for some rest. Starr will receive the best of care — you know that, don't you?"

Philea did not move, but something about her face caused the Healer to remove his hand from her shoulder very quickly — as if it were a dangerous location. A light flickered in her dark blue eyes, and she said, "The best of care for her body, yes. But there are other dangers."

The tall Healer shook his head, then bit his full lower lip. "There are fine Spiritual Healers on call, Philea. As soon as Starr wakes up, we'll see that they are available."

"No." Philea gave that monosyllabic answer, then forced herself to say, "Thank you for coming by."

"Certainly!" The Healer had lost some of his smooth assurance. His smile was forced and he was thinking *I'd like to take this crazy woman and wring her neck!* But he merely shrugged, saying, "Call me if you need me — anytime."

The two Healers left the room, and Philea sat down beside Starr. Her lips were dry and she was obviously at the end of her rope emotionally. "I'll wait until God does something," she murmured quietly.

Until God does something — but was God

going to do anything?

The thought fell into her mind, lay there like a stone — a tiny lump of doubt — like yeast in a loaf. And the longer she sat there struggling, the larger it seemed to grow.

Then the voices began.

Thoughts, really, but they came to her almost audibly. She knew it was the humming of the air conditioner, but the steady sound was overlaid with thoughts:

Maybe it's not God's will to bring her back. Maybe He's done this out of His sovereign will.

You're tired. You're not able to think or pray properly. Go home and get some rest — then you'll feel much better. You'll be able to pray better.

"No!" She forced her gritty lids open, opened the blue bag and tried to read the worn pages it contained — but her eyes were burning. She stared at Starr's face, and a bitterness rose in her, and anger boiled within:

It's not fair, God! I can't do it alone — why don't You send someone to help? — If she dies, it won't be my fault — You're asking too much!

If it had been a physical battle, that would have been easier — for although the pain of a fleshly wound could be fierce, it was as nothing compared to the warfare that had racked her spirit for so many hours. Her spirit seemed to be isolated in a cold, alien place, and all around were fierce, cruel beings that ripped and tore at her without ceasing. She thought of Prometheus chained to a rock and the vultures that came each day and tore at his liver, and she thought of the lone trapper lost in the woods surrounded by a pack of green-eyed slavering wolves, keeping them at bay by a single torch, knowing that if he dropped that small firebrand for one moment, the pack would be at his throat instantly.

She had fought this battle before, long ago in the Rim, but her father had been beside her. The two of them had stood against witchcraft and the black magic of the covens that had taken root even there — but now she alone kept one small flickering flame of faith held high. She alone knew of the darksome creatures, the powers and dominions that strained to move into Starr ripping her soul with their fell powers. As Moses' arms had grown weary held high to God, her spirit and her faith were dragged down as the long hours and days plodded by — and there was no Aaron or Hur to hold *her* hands high!

She struggled against the anger and bitterness, but it swept over her like a current, choking her attempts to pray. She had the sensation that she was being sucked down into a dark, resistless maelstrom, being drawn into a world of terror and fear. She uttered a small cry, and rose to her feet, tottered over to take the limp hand of Starr, and through parched lips quoted the ancient words: "Have mercy upon me, O Lord; for I am weak; My soul is also sore vexed; but thou, O Lord, how long?"

She fell across Starr's still form, like Elijah of old, and for a long time lay there sobbing and calling on God.

Finally she rose, wiped her eyes, then made her way to the bathroom taking the pitcher of ice from the table. Closing the door out of habit, she took a clean cloth and began bathing her face with cold water, then made an ice pack and held it over her burning eyes. She never was able to remember how long she stood there, but she did remember hearing the door close. The nurses and aides came in from time to time, and she thought nothing of it — but usually whoever came said a word to Starr, knowing that she could not hear, but out of habit.

But there was a silence in the room, and suddenly Philea dropped the ice pack and stood bolt upright — for the presence of something evil came to her, icy and deadly!

She whirled, threw open the door and saw instantly the head nurse bending over the still form of Starr, holding a pair of gleaming scissors in her hand, poised to snip a lock of

Starr's long hair! The nurse had always been cheerful, her round face honest and sincere.

But it was not the same, Philea saw instantly, for the simple face and honest eyes were transformed by something inside the woman that narrowed her eyes to slits and turned her lips into a snarl. At once Philea felt the same presence of evil power that she had felt years ago when she and her father had faced a painted witch doctor — for the eyes of the nurse had the same diabolical light that had burned in the eyes of the black magician.

"Leave this room!" Philea said instantly.

But the woman snarled in a voice not at all like her usual tone, "Get out while you can, old woman!" She dropped the lock of hair and reversed the scissors, holding the gleaming instrument like a dagger, and took a step toward Philea. "I'll kill you!"

Philea stood straight, looking into the woman's frenzied eyes, knowing that she stood on the brink of death — but there was no fear in her. She lifted her voice and cried out, "In the name of Jesus Christ of Nazareth, I rebuke you!"

The nurse faltered and the hand holding the scissors dropped slightly — but she shook her head and hissed, "You cannot stop us! Don't stand between me and what I've come for!"

Philea began to speak, and it was with a voice of authority, "In the name of Jesus Christ, I command you, evil spirit, begone! You have no power, Satan! The Lord Jesus Christ overcame you at the cross!" Then as the woman in front of her began to writhe and twist, her eyes burning, Philea cried out, "The blood of Jesus! The blood of Jesus Christ of Nazareth!"

The nurse dropped the scissors, covered her ears with her hands, and then ran out of the room making incomprehensible moans. The door slammed behind her, and suddenly Philea felt that every drop of strength had been drained from her. Her legs began to tremble and she staggered forward, falling on her knees beside Starr's bed. Her head swam, and she clutched at the sheets, fighting the weakness, knowing that she must not fall to the floor!

She felt herself slipping in a warm comfortable darkness, and cried out, "Lord, God! Strengthen me!"

And then — she felt a hand on hers!

A thrill raced along her nerves, for she thought for one panic-stricken instant that the demonic nurse had returned! But then she heard a thin, reedy voice:

"Mother! Mother! — where are you?"

She pulled herself to her feet, and saw that

Starr had found her hand and was clinging to it. "I'm with you, Starr!" she cried out. "I'm right here!"

The translucent eyelids of the girl on the bed fluttered, then opened wide, and Philea saw the panic in the girl's eyes. "Who are you?" she cried out, and Philea threw her arms around the girl, saying, "My name is Philea. I'm a friend. It's all right, Starr! It's all right! We've won! The victory is the Lord's on this day!"

She felt the girl's body move, and when she leaned back, she saw that tears were running down Starr's cheeks — but they were tears of joy! The eyes were clear, completely free from fear, and Starr whispered, "Philea — you brought me back! They were all trying to drag me down, but I saw this little light and I heard your voice — so I struggled to get to it." She looked at Philea with wonder in her eyes and said proudly, "You held them back! They were all horrible and black, coming across the bridge to drag me back with them — but you stood there and they couldn't get me — because of the light you held!"

Philea didn't understand the allusion to a bridge — but she knew that the girl had been saved, and she held her tight, rocking her back and forth, making a comforting noise. Finally Starr drew back and her eyes were huge as she said, "We won, didn't we, Philea?"

Philea nodded slowly. The glory of victory was in her pale blue eyes, but she said, "Yes — we won, Starr, but this was only one battle. The war will be long, and the enemy is strong. He will come again, and he is legion." But there was victory in every line of her face as she whispered, "But God is going to do a mighty work, Starr. He is going to raise up an army! It will be like no army you ever saw — and the world will laugh at it. But if we will be faithful to Jesus, we will see His banners flying in glorious triumph over this city!"

"An army, Philea?"

"Yes, Starr — and you are called to be a part of it!"

Chapter Nineteen

The Prophet

"Hurry, hurry! This way!" Lido waved frantically to David and Philea from the opposite end of the long corridor, glaring brightly under the florescent lights. He still felt uneasy at the sight of David in the uniform of a Peacemaker — although he had had twenty-four hours to get used to it.

David carried Starr, still dressed in her white hospital gown, in his arms as she clung to his neck, only half-conscious, her face almost the same color as the gown. Philea held to his right arm, dragging along behind him, weakened from three sleepless days and nights of fasting and prayer.

"Down here, in the Closed Ward! Show some life you lumbering idiots — they're escaping!" Bernard Alpha turned the corner just in time to see the four of them disappear into the storage room.

Lido led the way to the furthest corner of the cluttered room, piled high with boxes and crates of medical supplies and machinery.

Throwing aside a large white box, he bent down and slid open a panel at the bottom of the wall. Philea went through first, turning to help Starr as David lowered her from his arms.

The door burst open with a sudden light — the huge man in the white uniform pointing a shotgun into the room. He spotted David bent over in the far corner and raised the weapon to his shoulder. Lido scrambled to the top of a crate, a rubber tipped hammer with a shiny steel handle raised in his hand. He flung it at the Peacemaker with all his might, striking the hand that held the stock of the shotgun.

BOOM! BOOM! BOOM! The shotgun blasts shattered the quiet of the ward, but the hammer had deflected the barrel just enough, the heavy pellets striking four feet to David's right, ripping out sections of the wall. While the man reloaded, David threw off his helmet, reaching inside the opening where he had left his weapons. He strung the small crossbow, inserting a bolt as he spun to face his attacker. In one smooth motion, he swung the bow around, leveling it at his assailant as he squeezed the trigger.

The tall man had David in his sights when the iron-tipped bolt struck the front of his helmet, burying itself up to the feathers. A stream of blood gushed forth beneath the dark visor of the helmet just before he crumpled to the floor. Four other Peacemakers rushed over the fallen man as David followed Lido through the small opening, closing the panel behind him.

The darkness of the subterranean world beneath the City, sliced with pale striated light from above, enveloped the four fugitives like a damp, heavy blanket. Lido led them along a shelf of rough concrete that formed part of the foundation of the White Tower. David carried Starr with Philea bringing up the rear.

"Can they get to us down here before we make it back to the boats?" David breathed easily.

As if in answer to his question, a heavy door swung open fifty feet in front of them, the uniforms of the two Peacemakers shining in the glare of the florescent light as they stepped onto the concrete walkway at the base of the door. David saw with great relief that they both carried long slim swords rather than shotguns.

Easing Starr down beside Philea, David handed Lido the crossbow and bolts. The two men in white, their heavy muscles bulging under the shiny uniforms, advanced carefully toward David, swords held in front of them.

David glanced at the long iron staircase that descended from the concrete ledge at the base of the door, down into the twilight world where their boats awaited them. Twenty feet away now, the giants in white prepared to make their move. David unslung the oak staff from his back, balancing it easily in both hands.

Turning his head toward Lido, David whispered quickly, "When I finish here, take the girls down to the River. There may be more coming — I'll follow soon."

The first man had barely begun the thrust that would have plunged his blade through David's chest, when the oak staff whipped under his outstretched sword arm, shattering his right knee. Sudden pain paralyzed his whole body as the leg caved in, sending him tumbling down the steep concrete embankment that fell away from the ledge. His scream of agony rose up from the darkness far below.

With a roundhouse motion, the other man swung his sword at David's head, just missing as the blade clanged against the wall of the White Tower. At the same moment he heard the ringing of steel against concrete, his breath left him in a whooshing outrush as the butt of the staff struck his solar plexus. He doubled forward, his sword lost and sliding down the embankment.

Laying the side of his staff under the man's left shoulder, David flicked him after the clattering sword. "Down the stairs," David barked to Lido. "I'll meet you at the boats."

With Lido on one side and Philea on the other, they half-carried Starr down the long, steep staircase toward the River.

"Don't be long," Lido called back over his shoulder. "The Tower's full of those big bullies."

Stripping off the white uniform, David watched the three of them descend into the murky depths below the City. Through the open spaces between the concrete pillars and steel beams, he could see the wide clear channel of the River, thrown into shadow by the high banks, as it ran straight as an arrow shot toward the south, disappearing into the purple distance.

Hearing a noise behind him, David whirled to face another giant of a man rushing through the door. He lowered the barrel of the shotgun as soon as he spotted David, but coming out of the glaring light it took a fraction of a second too long. Before he could squeeze off a shot, David sliced the weapon from his hands with the staff, sending it rattling off into the darkness.

The man bellowed in pain as the polished

oak cracked against the knuckles of his left hand. In a blind rage, he got his huge hands around David's neck, squeezing with a strength born of desperation and a hatred of anyone who dared defy his masters.

David felt himself slipping quickly into a smothering blackness under the viselike grip of the grunting giant. Holding to his attacker's arms for balance, he stomped the heel of his right foot down on the man's left instep, crushing it with a sound like a limb snapping underwater. Releasing David, the man screamed in pain as he fell to his knees. David slammed his right fist into the man's temple. He rolled over on his side, lying perfectly still.

Hearing the muted thunder of dozens of footsteps rushing down the hallway toward the open door, David looked quickly about him. Taking his staff from the ledge where it had fallen, he put his shoulder to the door and closed it, wedging his staff between the heavy metal door and the corner post of the railing. A tremendous pounding came against the door with the sound of shouting as the pursuers tried to break it down. With a brief look back to make sure the staff was secure, David bounded down the stairs, taking them three at a time.

"You go ahead, Philea. I'll try to help Starr

across after you." Lido stood on the east side of the River, easing Starr into a sitting position on a pipe that paralleled its bank.

"You'll never make it alone," Philea insisted, breathing heavily after their long descent.

Lido glanced at the eight-inch wide I-beam that spanned the hundred-foot width of the River. "We'll make it," he muttered with feigned confidence.

Philea reluctantly stepped out onto the cold surface of the beam. Almost immediately, a half dozen Rippers appeared in the River, as if by magic, along the route she would take over the water.

How do they know? How do they always know when someone is over the water? Lido stared with a mounting dread at the spiny, gray shadows, circling ominously beneath the surface, their eyes reflecting the dim light like cold fires.

Looking down and taking a deep breath, Philea began the walk across. Slowly and carefully, her arms outstretched for balance, she gained the opposite bank.

Lido struggled to get Starr to a standing position.

"You next!" David leaped down the last of the stairs, sprinting across the open space to the River bank. "Go ahead — don't stand there! I'll bring Starr."

"Thank God, you're all right!" Lido declared excitedly. "Are they coming?"

"They're coming," David shot back, glancing over his shoulder. "It's just a matter of how long we have."

Lido scurried across the beam in a flash, running over to Philea to help her down the ladder to the boats.

Lifting Starr like a child in his arms, David crossed the narrow bridge and climbed easily down the ladder, where Philea and Lido were already in their boat. David laid Starr gently in the small bow of the boat, climbing in carefully and paddling strongly out into the current ahead of his companions. Below the fragile crafts, the Rippers glided like the antithesis of guardian angels.

David kept to the middle of the River so they would be as far as possible from a threat on either bank. Moving with the current, they passed swiftly under lofty girders and beams and past the huge concrete pillars. The anemic light from the high vaulted ceiling glinted like slivers of ice in the depths of the River and off the armour plated backs of the Rippers.

Glancing ahead and upward David noticed movement in the shadowy heights. A small man, bald and wearing a light brown tunic, stepped out onto a girder a hundred feet above the River. David recognized him as one of the men who had visited Starr in the Closed Ward. Almost instantly, a Peacemaker appeared next to him, handing him a shotgun. The small man took the gun, jacked a shell into the chamber, and aimed it directly at Starr's prone form. David leaned forward, gripping the gunwales for balance as he shielded Starr's body with his own. He felt fire in his left shoulder as two of the pellets ripped through the fleshy part at the edge of his ribcage. The hollow roar of the gun reverberated through the high framework like a distant cannon.

Bernard Alpha had miscalculated the distance, but seeing that his victims were helpless, he pumped a fresh round into the chamber and took deliberate aim. He heard a soft thunk — his right thigh exploded with a scalding pain as the bolt from the crossbow buried itself into the soft flesh. Alpha's leg collapsed under him. Losing his balance, he clutched at the dark air around him. With a piercing wail of despair, he plummeted toward the River. The last thing he saw before he plunged beneath the surface, were the gleaming yellow eyes of the Rippers.

David glanced back at Lido, who lowered the crossbow and inserted another bolt. In front of David where Bernard Alpha had hit, the water appeared to be boiling, changing color quickly from a deep pinkish tint to bright red.

The Peacemaker, seeing Lido reload the crossbow, stepped behind a wide steel brace, bringing his shotgun to bear on the tiny man in the second boat. As he tightened his finger on the trigger a small shadowy figure swung from a rope tied to a support high above him, slamming his feet into the big man's back. The shotgun thundered as the man, arms and legs flailing, pitched forward down to the surface of the River. Again the clear water became a boiling, crimson cauldron.

The Shadowman, still holding his rope, stood where the Peacemaker had been; a bright smile on his dark face. When Lido waved his thanks, the man tipped his greasy cap, disappearing like a vapor back into the dim recesses of his home under the City.

"They'll blanket the City with searchers. We've got to get out now." Michael Kappa paced back and forth in his cramped quarters, a restless light in his eyes.

Starr sat next to David on Michael's bunk. Her pale, almost ghostly face reflected the horror of the past few days. A streak of pure white ran through her thick auburn hair from her left temple along the side of her head to her shoulder. She smiled up at David, grateful that his wound, now cleaned and bandaged, had not been serious. Leaning back against his chest, she allowed herself the luxury of relying on his strength rather than her own.

David put his arm around Starr, feeling her fragile softness against him. "Starr's in no condition to travel — neither are you for that matter, Philea, but Michael's right. We've got to get across the River and through that laser gate before they block off everything."

Lido and Philea walked over next to the others from their temporary seclusion in the corner. Philea placed her hand on Lido's shoulder, a soft light in her eyes. "We've decided to stay."

Michael spun toward them. "Why — that's insane. They'll hunt you down like animals."

Lido returned Michael's look with a newly found confidence. "We're going underground — to live with the Shadowmen and the other Christians who've had to escape there."

Michael opened his mouth to respond, but was waved gently to silence by Philea.

Sitting next to Starr, Philea smiled at each of her four companions. "The dark night of the soul is come upon us all, my friends. It will begin here in the City, but I can see now that it will also extend to the Fields and wherever else Christians may be."

"We can escape them in the Fields, Philea!" David offered fervently. "It's the one place where they're at a disadvantage — even with all their weapons. They're soft from all the easy living and having no one in the City stand up to them. The Border Guards are our only real threat and from what Michael tells me, a lot of them won't try very hard to hunt us down."

Philea gazed intently into David's eyes. "They would be at a disadvantage, David — if they played by the same rules you do — but they don't. They'll try to destroy you by hurting the weakest and least suspecting of your people. They'll come against you with no sense of honor."

"Do you really think you'll be safe down there under the City?" David asked gravely.

"Lido does," Philea smiled. "Anyway, we must stay. Our people will need us more than ever now. Not just in ways to help them escape the inquisition — but most of all with prayer. It's the most powerful weapon we have — the only way we'll be victorious."

As Philea spoke, a sudden, chilling picture filled Starr's mind. She saw herself walking down the corridor between the high stacked cylinders of the Rim toward Martha Epsilon's home in number seventy-three. A pale woman in her mid twenties sat above her at the edge of cylinder number forty-nine. She had blonde hair tied at the back with a piece of twine and her light blue eyes were fixed on a picture she held in both hands.

Starr felt soft hands on her head and, glancing upward, saw that Philea was praying for her — for God to give his angels charge over her to protect her in all her ways. She laid hands on each of the other three in turn, praying for God's grace and protection over all of them. When she had finished, she sat down next to Starr, as if the last of her strength had been exhausted.

Taking Philea's hand, Starr felt hot tears flow down her cheeks as the horror of what she had done rushed in upon her. The three men left the room to make final preparations for the trip as if bidden to do so by someone unseen.

Overwhelmed now by grief, Starr was sobbing, her body shaking as she gasped for breath.

Philea put her arm around Starr's shoulder, letting her cry herself empty. "It's all right, my child," she murmured, her voice no more than a whisper in the still room. "It's all right now."

Although they were the same age, Starr felt

that Philea was somehow much older than her although she had the face of a child. "But, Philea, you don't know what I've done."

Philea touched Starr's cheek with her hand, turning her face toward her.

Starr looked into Philea's blue eyes and saw a love and an understanding that were beyond her comprehension. "Philea! " she whispered, turning away. "I murdered your father the same day I — went to Martha Epsilon's. Oh, please forgive me — forgive me!"

Philea took Starr's hand again. "I already have, my child — that same day."

Starr gazed at Philea, her eyes wide in disbelief. "You — you already knew what I had done! And you risked your life in the White Tower to save me!"

"I was only able to forgive you by the grace and love of Jesus Christ, Starr," Philea said in a steady voice. "I had to fight the battle in His strength that day, for I knew the bitterness would destroy me if I didn't. By His mercy, it's all gone."

Starr put her arms around Philea, and as they embraced she felt like warm oil was flowing through her, cleansing and healing all the old wounds and hurts.

Sitting back, Starr looked again into Philea's eyes. "I — I haven't told David what I've done — what I was. I just couldn't do it."

"The right time will come," Philea assured her. "God has forgiven you and David will too."

Lido slipped quietly into the room. "They're ready."

Starr looked down at the little man with the big bright eyes. "Thank you so much for what you've done. I do so wish you were going with us." She bent down and kissed him on the cheek.

"I want to see the Fields very much," Lido murmured, smiling up at Philea. "But we've got work to do here. Maybe we can both come to all of you later."

"Oh, I hope so," Starr said, turning to Philea. "I'll pray for both of you every day."

"And I you — and for all of those in the Fields," Philea replied as they embraced. She stepped back, holding Starr's hands and gazing intently into her eyes. "God will send a special anointing upon you, Starr Omega. He has chosen you among all His other children for a particular work. Go now in His grace."

Starr squeezed Philea's hands gently, feeling the tears come to her eyes as she turned and left the room. As she hurried toward David who waited at the foot of the ladder, it occurred to her that perhaps it was not by accident that this healing and cleansing of her spirit had happened in a stable.

As they crossed the small river that bordered his home — further north where it joined the much larger one, David saw the huge column of smoke. Rising against the hard blue sky it became an ominous harbinger of the wolf-lean years ahead, bringing to mind with a chilling finality Philea's words spoken in the stable, "They'll come against you with no sense of honor."

David rode ahead along the path through the trees next to the bluff that looked down on the River — Starr followed on her Palomino with Michael acting as rear guard. As they came to the vegetable gardens, David saw the first of the bodies. Women and children had been hacked to pieces, thrown carelessly among the carefully tended rows like so much garbage.

On the opposite side of the path from the gardens, the livestock had been slaughtered and lay scattered among vast pools of blood. Smoke eddyed from the burned pens, wafting upward as the breeze off the River caught it.

The three travelers looked straight ahead as they continued down the path toward Haven, dreading what they would find there. The heavy drone of hundreds of blue-bottle flies overlay the grisly scene like some ghastly soundtrack.

"There's where the gunships landed," Michael remarked, struggling to keep his voice steady.

Starr looked at the open field where the long grass had been flattened by the helicopters. She saw David staring at something beyond the field next to the pond. His face paled suddenly, as he kicked his horse into motion, racing through the tall grass at breakneck speed. She and Michael followed at a gallop, knowing that speed no longer mattered in this holocaust called Haven.

When Starr climbed down off her horse, David sat on the hard-packed ground under the ancient sweet gum tree, cradling in his arms the body of his butchered mother. Timmy's small lifeless form lay next to him, where David had placed Timmy's head on his knee as though he were asleep.

Together they walked up the slope toward the charred remains of the house, David carrying his mother and Starr carrying the child. Michael followed, leading the horses. He was glad to no longer be a servant of beasts calling themselves human — those who could commit such atrocities, even if it meant death at their hands.

Near the house, under a huge oak blackened by the fires, Obadiah sat next to the body of David's father. His face and chest had been ruined by shotgun blasts.

Obadiah looked up at the three travelers, recognizing two of them. His face was slack, the eyes glazed by grief and shock. "Your daddy fought as hard as he could to protect his family, David," he muttered, pointing to the bloody scythe still clutched in Caleb's work-hardened hands. "But I reckon he didn't stand much of a chance."

David laid his mother's body next to his father's. Starr laid Timmy between them.

As David took Obadiah's hand, helping him to his feet, he looked about the village where several men, just back from their time in the fields, were wandering aimlessly amid the smoldering ruins. "Come on, old friend," David said softly. "We've got work to do."

Late that night, their grisly work completed, the survivors sat around a fire under the tall pines on the bluff that overlooked the River. The flames crackled with a friendly warmth, the night wind sighing high in the crowns of the trees. Beyond the edge of the firelight mounds of fresh earth stretched in orderly rows.

"Light of the World!" Obadiah spoke softly, but with an unmistakable authority. He rose to his feet, this small black man wearing a rough homespun cloak. His face seemed transformed, shining with a strength eternal and immutable.

Starr moved closer to David, holding to his arm as Obadiah spoke. The others lifted their faces in rapt attention at the compelling sound of his voice.

"I have a word from the Lord," Obadiah continued. "We are the Light of the World and that Blessed Light must not be darkened by thoughts of revenge against those who have slain our loved ones. 'Recompense to no man evil for evil. Dearly beloved, avenge not yourselves, but rather give place unto wrath: for it is written, Vengeance is mine; I will repay, saith the Lord.' "

Obadiah paced around the fire. Looking into the upturned faces, "Jesus said that we would be hated by all men for his sake. Some of us will be betrayed by those closest to us: parents, brethren, friends — and some of us will be put to death. There will be great earthquakes and famines and pestilences — fearful sights and great signs from heaven — signs in the sun and in the moon and in the stars."

A smile of great joy came to Obadiah's shining face. " 'And when these things begin to come to pass, then lift up your heads; for your redemption draweth nigh.' Our redemption is very near, my dear brethren and my sister. Jesus is coming soon. Some of us will be witnesses for him here in the Fields — others will journey to the City where you may be called upon to preach the Gospel to those who would murder you."

"Brethern, let us love one another. Let us rejoice in the grace and mercy of Almighty God. Paul said that he had suffered the loss of all things that he might know Jesus — 'and the power of his resurrection, and the fellowship of his sufferings . . . For I reckon that the sufferings of this present time are not worthy to be compared with the glory which shall be revealed in us.' "

Obadiah lifted his arms toward the great black dome of heaven sparkling with thousands of stars and the shining planets. "Father, into your hands we commend our lives — our hearts — our spirits. Jesus prayed for us not to be taken out of the world, but that You would keep us from the evil. We trust in this prayer and in your eternal love for us which is in your son, Jesus."

With a smile that would not allow fear in its presence, Obadiah turned again to the remnant gathered around the fire. "I leave you now with the words of Jesus, 'Fear not, little flock; for it is your father's good pleasure to give you the kingdom.' "

Sitting down at the foot of a pine, Obadiah, a small black man in rough clothes, insignifi-

cant and of no value in the eyes of the world, lay back and closed his eyes.

As the little group settled down for the night, Starr lay next to David on a bed of fragrant pine needles, her hand securely clasped in his. She thought of the great and terrible things that were to come, but strangely, there was no fear in her heart — there was only peace and joy and a deep longing to see the face of Jesus.

She imagined Jesus in the hills of Galilee with his rough band of followers, gathered around a fire such as this. She could almost see him standing before them — could almost hear his voice as she drifted off to sleep.

Fear not, little flock.